Dragonflies and Damselflies of Oregon

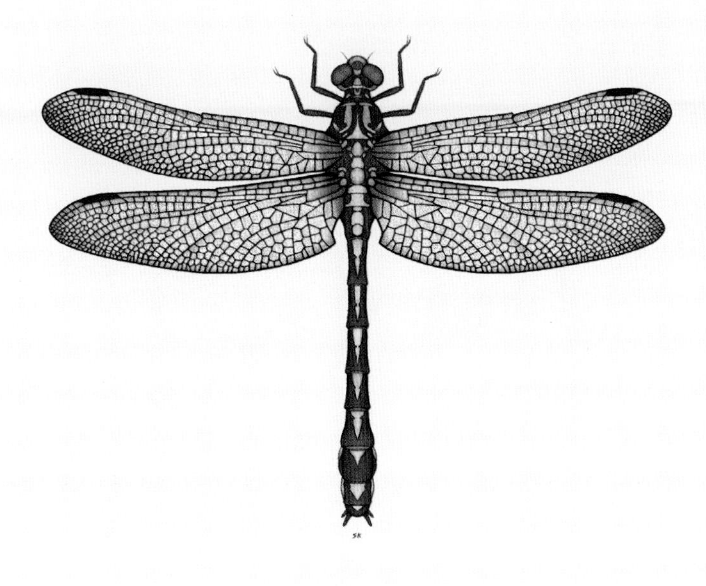

The John and Shirley Byrne Fund for Books on Nature and the Environment provides generous support that helps make publication of this and other Oregon State University Press books possible.

Previously published with the support of this fund:

One City's Wilderness: Portland's Forest Park
by Marcy Cottrell Houle

Among Penguins: A Bird Man in Antarctica
by Noah Strycker

Additional financial support for publication of this guide was provided by an anonymous donor through the Oregon Community Foundation.

Dragonflies and Damselflies of Oregon

A Field Guide

Cary Kerst and Steve Gordon

Oregon State University Press
Corvallis

The paper in this book meets the guidelines for permanence and durability of the Committee on Production Guidelines for Book Longevity of the Council on Library Resources and the minimum requirements of the American National Standard for Permanence of Paper for Printed Library Materials Z39.48-1984.

Library of Congress Cataloging-in-Publication Data

Kerst, Cary, 1945-

Dragonflies and damselflies of Oregon : a field guide / Cary Kerst and Steve Gordon.

p. cm.

Includes bibliographical references and index.

ISBN 978-0-87071-589-1 (alk. paper)

1. Dragonflies--Oregon--Identification. 2. Damselflies--Oregon--Identification. I. Gordon, Steve (Steven), 1947- II. Title.

QL520.2.O7K47 2011

595.7'33--dc22

2010053133

Printed in China

Oregon State University Press
121 The Valley Library
Corvallis OR 97331-4501
541-737-3166 • fax 541-737-3170
http://oregonstate.edu/dept/press

To my Grandmother, Victoria Ann (Deno) Senesac (1882–1977), who always had a kind word and a treat for any of her 152 grandchildren and great-grandchildren who came to visit. She was a woman of remarkable strength and faith and the center of our large family.

Cary Kerst

I dedicate this book to my family. My wife and friend for the past forty-six years, Susan, has contributed, been supportive, and given good advice. My children, Josef and Kimberly, love Oregon and nature. My son-in-law, David, and I enjoy our birding adventures and walks. And now Kimberly's and David's children, Benjamin and Emma, my grandchildren, provide opportunities for me to relive the wonderment of the natural world. To them belongs the future.

Steve Gordon

We also dedicate this book to the land owners, land managers, conservationists, environmentalists, educators, and citizen volunteers who protect and restore Oregon's watersheds—its lands, open spaces, rivers, streams, lakes, ponds, and wetlands. These waters and surrounding environs are critical habitat for aquatic life, including our wonderful Odonates, Oregon's dragonflies and damselflies.

Cary Kerst and Steve Gordon

Table of Contents

List of Figures

List of Maps

Individual range maps for Oregon's 91 Odonate species are located within each species' description pages.

List of Identification Charts

We offer our thanks to the anonymous donor who provided financial support through the Oregon Community Foundation to Oregon State University Press for publication of this guide. This support helped make this guide possible.

We want to thank Jim Johnson and Steve Valley for sharing their expertise on the Odonata of Oregon. We have benefited from their generosity in sharing their knowledge of the Odonata and Oregon dragonflying spots. We look forward to their book on the Odonata of the Pacific Northwest. We also thank all of the members of the Oregon Dragonfly Survey for comradeship, wonderful days in the field, and evenings around the campfire. You all help us recapture the wonder of the natural world.

We thank Charlie Quinn for his enthusiasm and encouragement over the years. Our gratitude is also due for review of the manuscript from a naturalist's perspective.

The wonderful maps in this guide were made possible through the work of Joe Gordon and Jeff Krueger who continued to make change after change for us. Stephanie Korshun provided her color drawing of the Columbia Clubtail on the title page.

At Oregon State University Press, we thank Mary Braun, Jo Alexander, and Judy Radovsky for their work and encouragement in making this guide a reality. The work of Oregon State University Press provides an important service to the citizens of Oregon.

We thank the following exceptional photographers for allowing us to use their photos: Giff Beaton, Steve Berliner, Ryan Brady, Ray Bruun, Jim Johnson, Ron Oriti, Dennis Paulson, Ken Tennessen, and Steve Valley.

Photos

Giff Beaton: emerald larva (*Corduliidae*), spiketail larva (*Cordulegastridae*)

Steve Berliner: photo of Steve Valley

Ray Bruun: River Bluet females (*Enallagma anna*), Northern Bluet female (*Enallagma annexum*)

Ryan Brady (pbase.com/rbrady): Ocellated Emerald female (*Somatochlora minor)*

Steve Gordon: Desert Whitetail immature male (*Plathemis subornata*), Illinois River, Magone Lake, Mickey Hot Springs, Todd Lake.

Jim Johnson: Black-tipped Darner female (*Aeshna tuberculifera*), Shadow Darner female (*Aeshna umbrosa*), Vivid Dancer female (*Argia vivida*)

Ron Oriti: Olive Clubtail female (*Stylurus olivaceus)*

Dennis Paulson: cruiser larva (*Macromiidae*), Red Rock Skimmer female (*Paltothemis lineatipes)*, Familiar Bluet female *(Enallagma civile)*

Ken Tennessen: Cherry-faced Meadowhawk female (*Sympetrum internum*)

Steve Valley: Oregon Odonata Survey group photo

All other photos

Cary Kerst

Title page illustration

Stephanie Korshun: Columbia Clubtail

All other illustrations

Steve Gordon

I. Introduction

You've seen colorful insects with their bold, bright colors performing high-speed aerial acrobatics at a pond or stream. Perhaps you have even seen them flying or perched in your yard or garden. The size and colors of these large insects naturally draw our attention and curiosity. They are the dragonflies and damselflies, two suborders (Anisoptera and Zygoptera) in the insect order Odonata, collectively referred to as Odonates. Anisoptera comes from Greek meaning "unequal wings" and refers to the different shape of the forewings and hindwings in the dragonflies. Zygoptera has a similar derivation meaning "joined wings" and refers to the similarly shaped front and hind wings.

The Wonder of Dragonflies

This field guide presents the science, photos, and illustrations to assist with the identification of Odonate species. We also, throughout the book, present photos of dragonflies and damselflies simply meant to highlight the beauty and visual impact of these amazing creatures. Recapture the wonder!

Black Petaltail on California Pitcher Plant

The name Odonata is derived from the Greek referring to the toothed mandible of these insects. These hardy insects are predatory carnivores. Collectively, they consume millions of insects daily from gnats and mosquitoes to bees, butterflies, and even other Odonates! Worldwide, there are 5,740 species of dragonflies and damselflies with 457 species found in North America (Paulson and Dunkel, 2009). There are currently 91 species known from Oregon. New species are yet being discovered, especially in the tropics.

Dragonflies and damselflies have become very popular subjects for observation, and interest in this fascinating insect order continues to grow. Entomologists and fishermen have long been attracted to the Odonates, but now bird and butterfly enthusiasts have found them to be worthy subjects. Gardeners have come to appreciate them for their beauty but also for their voracious appetite. Many nature photographers are also focusing on the Odonates. While they are most often associated with water, the adults range far and wide. They begin their interesting life as eggs, passing to aquatic larvae, then emerging to become flying insects. This transformation from a life underwater to flying above our heads is truly amazing. Both the larvae and adults are beneficial, consuming large numbers of other invertebrates—many of which people consider pests.

You will find adjectives in this field guide attempting to accurately describe these insects: colorful, magnificent, iridescent, beautiful, dazzling, wonderful, bright, spectacular, and more. As you become familiar with more species, you will discover that no adjective can do justice to a dynamic, living dragonfly or damselfly.

There are numerous myths and beliefs about this order of insects, and many of us have heard the folk names that hint at some of these: names such as *devil's darning needle*, *horse stinger*, and *snake feeder*. In fact, the Odonates are harmless creatures—unless you are another insect, of course!

The Dragonflies and Damselflies are an ancient order of insects with a fossil record now going back about 425 million years. Permian fossils record a dragonfly with a 27-inch wingspan. It is fascinating to imagine what such an insect ate and the other life forms on earth at that time!

When we began studying Odonates with the goal of making a checklist of species in our local area, there were limited field guides available. Due to increasing public interest, the number of guides being published has increased. This guide is the first one specific to Oregon. Our goal is to provide a resource that will promote interest in the Odonates and allow you to identify the 91 species found across this diverse state. From valley to mountain top, from ocean to desert, from river to lake, from city to remote country, you will find dragonflies.

To learn more about dragonflies, you will be drawn to water. That is a good thing. Once at the water's edge, whether on the shore or knee-deep in the shallows, you will hear the flutter and whir of dragonfly wings. You will see astounding aerodynamic displays taking place at speeds that will challenge your eyes' ability to keep pace. You will observe unique Odonate mating displays and behaviors that put contortionists to shame. You will be amazed at the dramatic chases between competing males seeking females or defending a territory. These often take dragonflies so high into the sky that your eyes can no longer follow them. You will marvel at females tapping the water surface time and again as they deposit tiny eggs. If you ever lost it, you can regain the curiosity about the small things in the universe that is the domain of every child.

To study Odonates is to be introduced to a rainbow of colors. You will observe reds, oranges, yellows, greens, blues, lavenders, pinks, whites, blacks, grays, and browns in hues and patterns that do not seem real. The bright blues of the tiniest damselfly are truly electrifying. You may wonder at the purpose for such diversity.

We also hope this field guide promotes increased interest in the natural world and in the preservation of our aquatic habitats. There are still tremendous opportunities for those interested in Odonates to add to our knowledge about species' distribution, habitat preferences, life cycles, and behaviors. Another goal of this field guide is to promote greater involvement among naturalists in contributing to Odonate science. We have provided both the scientific name as well as common names in the text. We use common names as well as common terms for morphology as much as possible. We cover both male and females of each species. Enjoy your adventures along Oregon's streams, ponds, lakes, and wetlands.

Life History

Odonates go through three distinctive life stages: egg, larva, and adult. Each stage is described below and as shown in Figure 1.

The Egg

While eggs may begin hatching in a few days or weeks, some species are able to pass unfavorable seasonal or drought periods in the egg stage. When eggs hatch, the initial stage is the prolarva which only lasts a few minutes before molting. Since some eggs are laid in plant tissue, mud, or a gelatinous mass, this stage functions to allow the larva to escape from the egg's surroundings or, in some cases, to find water (Corbet, 1999).

The Larva

Favorite habitats for larvae include vegetated underwater shorelines and submerged logs. The larvae of darners (Family Aeshnidae) and damselflies cling to plants and logs, while some groups such as the clubtails (Family Gomphidae) and spiketails (Family Cordulegastridae) burrow in sediment. Larvae of the skimmers sit on surfaces such as substrate and vegetation. Larvae are voracious feeders, actively stalk prey, and larvae of large dragonflies such as darners will even catch small fish. Studies of several species demonstrated that larvae would consume 21-47 mosquito larvae in 40-50 minutes before ceasing to feed (Corbet, 1999).

Larvae have a hinged lower jaw (modified lower lip or labium) which can shoot out rapidly to catch prey. The movement of the lower jaw is so fast that it is usually not visible to an observer. This prehensile labium makes the larvae very effective predators. Dragonfly larvae breathe through gills located in a rectal chamber but may also be able to absorb oxygen through the abdominal exoskeleton. Water can also be rapidly expelled from the rectal chamber to allow larvae to escape predators, making them truly jet propelled for short distances. Damselfly larvae breathe through leaf-like gills that extend from the rear of the abdomen, as well as through the exoskeleton. Larvae have a thin chitinized exoskeleton (similar to the material in a fingernail), which must be shed as growth proceeds. The number of times the larva sheds its skin averages about twelve, but varies not only between species but within a species. After molting, the larva is more vulnerable to predators until the exoskeleton hardens. In addition to being effective predators, Odonate larvae are important components of the diet of other organisms such as fish, birds, turtles, and snakes (Corbet, 1999). We have included a photo of a typical larva from each family in order to illustrate the range of variation.

Emergence

The mature larva crawls from the water on emergent vegetation, rocks, or wood. Some larvae, such as those of the clubtail family which are burrowers, will be found on rocks or the banks along water bodies. Those that are good climbers, such as the darners, will be found on emergent vegetation, bridge abutments, or even trees. Emerging Odonates will often be found hanging vertically or even at a sharper angle, allowing the wings and abdomen to hang until expanded and hardened.

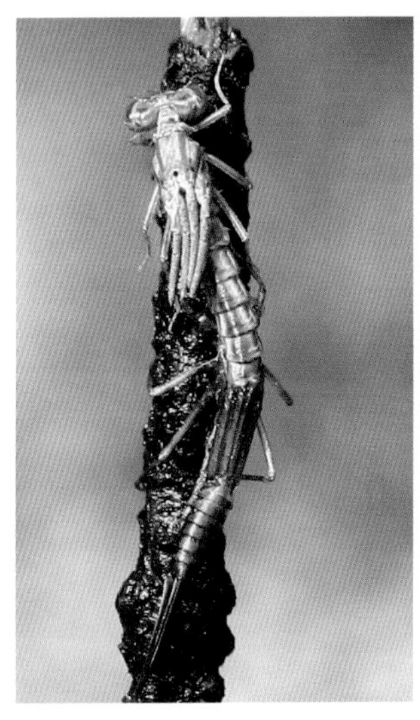

Damselfly emergence

Once out of the water, the exoskeleton splits along the back of the thorax, and the adult emerges. Blood pressure inflates the wings. The newly emerged adult is fragile, lighter in color than the mature adult, and reluctant to fly. These Odonates in the immature stage are referred to as "tenerals." Color develops with time, and mature colors develop more quickly at warmer temperatures. Males are more brightly colored than females in most cases. Females may also have several color forms. These multiple color forms make identification of females more difficult! The female may be colored in brown or green or colored like the male. The females that are colored like the male may be referred to as andromorph (male-form) or androchrome (male-color) while the dissimilar females are called gynomorph (female-form) or gynochrome (female-color).

Some colors, such as the blues, may darken with cool temperatures to grays and purples. For example, darners will often have darker colors until they warm up in the morning.

Perching in the sun in cool weather and exposing as much surface area to the sun as possible is a method used to warm up. Dragonflies will also perform "wing-whirring," a rapid movement of the wings to increase body temperatures, much as humans do by running in place. On extremely hot days, some species such as the Blue Dasher and Cardinal Meadowhawk will "obelisk" while perched. In this position, the abdomen is held near vertical pointing towards the sun. This reduces the surface area exposed to the sun and helps prevent overheating.

Blue Dasher obelisking

EGG LAYING →

Oviposition takes place in tandem or alone, depending on the species. Eggs are fertilized as they pass through the oviduct. Females may lay from a few hundred to thousands of eggs.

Cardinal meadowhawk pair

MATING

Belted Whiteface pair

LIFE

ADULT PAIR ←

During mating, the male grasps the female behind the head. The female bends her abdomen up to the second abdominal segment of the male. Sperm transfer occurs at this point.

Boreal Bluet pair

→ EGGS

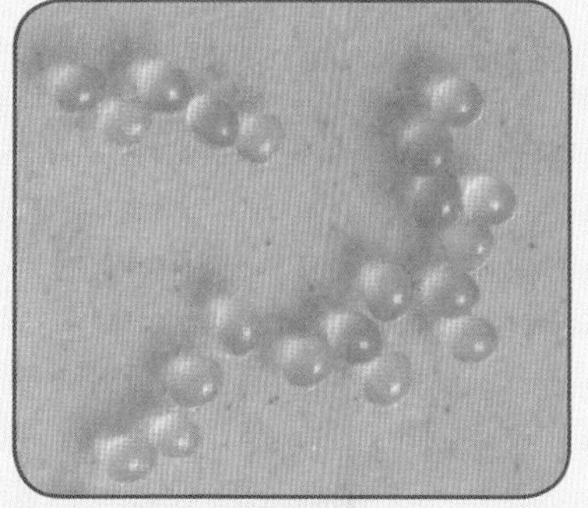

Striped Meadowhawk eggs (size: ~ 0.5 mm)

Eggs normally hatch in a couple of weeks, but may have delayed development in dry and/or cold conditions. Eggs may overwinter before hatching.

↓

LARVA

CYCLE

Darner larva

↓

← EMERGENCE

Variable Darner female with larval skin

Odonates molt 17–24 times during the larval stage, which lasts from a few months to 5 years. The larva crawls out of the water when mature and emergence to the adult takes place.

FIGURE 1

The Adult

Adult Odonates have large compound eyes capable of seeing movement very well and making them difficult to catch—for humans as well as predators! They breathe through spiracles (tubes connected to openings on the side of the body), feed on insects, and are important predators of pests such as mosquitoes. Dragonflies can often be seen darting in virtually any direction to catch flying insects when feeding. The aerial acrobatics that they perform are truly amazing. They will often fly straight up until lost to sight and are capable of speeds of 35 miles per hour. At times, dragonflies forage far from water while feeding on insects. The adults typically live from two to four weeks, but there are exceptions and some may live for months. Thus, one can see that the majority of the typical Odonate's life span is spent as a larva in water. The adult Odonate life is all about feeding and reproduction.

It is not uncommon to find adult Odonates with small red parasitic mites attached. The larval mite locates the mature larva and finds a hiding place and stays with the larva as it leaves the water. During emergence, the mite transfers to the adult where it attaches to feed on body fluids. Once fully mature, the mite drops off.

Studies indicate that species that perch and catch prey, as the flycatchers of the bird world do, will consume about ten to fifteen percent of their body weight each day in prey. Species that spend a lot of time flying must consume greater quantities to support this level of activity. The success rate for catching prey is variable but can be quite high measured at 51 percent for a species of percher (Corbet, 1999). For those of you who are baseball fans, think of this as a .510 batting average. Amazing! This is an impressive performance—especially given that prey are caught with the mouth—showing the great visual acuity of Odonates. Those species that hunt on the wing have lower success rates for capturing prey than the perchers.

Pacific Clubtails, male to male to female

Some dragonfly species patrol a territory along a pond or stream seeking mates and challenging other males entering the area. In fact, one may see dragonflies make passes at other animals in the line of vision, such as birds. Indeed, on a warm day, it sometimes seems as if every incursion is challenged or investigated. Adult males have abdominal appendages on the terminal segment called claspers or *cerci* that are used to grasp the female behind the head during mating. The male has secondary sexual appendages under the second abdominal segment. This segment is often enlarged, making it simple to determine male from female. In side view, there is an obvious protrusion on the bottom of the abdomen where the opening of the male's secondary sexual appendage is located. Under microscopic examination, the secondary sexual appendage is structurally very complex. Sperm are transferred from sexual organs in the male's terminal segment to this appendage just prior to mating. Odonates are often seen in a "wheel" position with the male grasping the female behind the head and the female bending her abdomen up to the second segment of the male's abdomen, where sperm transfer actually occurs, for mating.

The male's secondary sexual appendage may also be capable of removing sperm from the female from prior matings. Unusual pairings of Odonates have been documented, and we have made observations of this phenomenon. We have noted pairings of males with females of a different species as well as male to male to female connections. It is quite a sight to see three large dragonflies connected and flying in an erratic, high-speed flight.

Egg-laying behavior varies widely among species. Fertilization of the egg occurs as eggs pass through the oviduct during oviposition. In some species, such as skimmers (Family Libellulidae), the female can be seen dipping her abdomen repeatedly into the water to drop eggs.

Skimmer males may be seen guarding females during oviposition to prevent other males from mating with the female. Darner females can often be seen sitting on emergent vegetation or partially submerged logs, inserting eggs into plant tissue both above and below the water surface. Male darners may remain attached to the female during oviposition. In some species, such as the meadowhawks, the male continues to grasp the female behind the head to guard her against mating with other males. In this tandem position, they continually dip to the water surface, but some meadowhawk species are also commonly seen depositing eggs in grassy areas along ponds or in seasonal wetlands that will be submerged later by winter rains.

Life cycles vary from one year or less in many species to two or three years for the large darners. Some species, such as the Black Petaltail and Pacific Spiketail, require five years to complete a life cycle. There may be several generations per year in tropical areas. A single female may lay hundreds or thousands of eggs. We once observed a pair of Cardinal Meadowhawks hovering in the tandem position while the female dipped the tip of her abdomen 82 times above a mass of submerged vegetation. The damselfly male usually remains attached to the female during oviposition. They are easily observed landing on emergent vegetation or floating algal mats where females insert eggs into plant tissue below the surface.

On a warm summer day, it may seem like a chaotic frenzy as Odonates challenge each other, drive off competitors, defend territories, and oviposit. However there are many interesting behaviors that can be observed. We encourage you to make your own observations of the fascinating behavior of Odonates while in the field. Some species like the darners seem to fly during the day without ever perching. When they do perch, they usually hang on vegetation or the bark of a tree in a vertical position ("hanging up"). Some species such as the Grappletail and snaketails perch on vegetation, stones, logs, or the ground in a horizontal position. The darners hunt like swifts on the wing seldom resting until nighttime. Others, like the meadowhawks, hunt more like flycatchers—returning to the same perch after sallying out to nab flying prey.

Dragonfly migrations have been documented on the east coast of the United States. Migration of meadowhawks also occurs on the Oregon Coast, but this phenomenon is not well studied or understood due to difficulty of documenting and following their movement. These mass movements seem to be initiated by fall cold fronts.

Many Odonates are taken by predators. We have observed birds targeting Odonates where there were large concentrations of them. Odonates also will often be found in spider webs near water. We have watched a hornet attack a perching meadowhawk, sting it, and carry off the thorax after cutting it out. The remainder was then carried away by ants! "Eat or be eaten" seems to be the maxim of the insect world.

Robin with adult Grappletail pair

Morphology

The general morphological features of insects are shared by the Odonata. We use the accepted symbols for male (♂) and female (♀) in figures and charts. The body is divided into the head, thorax, and abdomen as illustrated in Figure 2A. The large compound eyes lie on each side of the head. The thorax is divided into three segments: the prothorax, mesothorax, and metathorax each bearing a pair of legs. The two large segments of the legs are the femur (nearest the thorax) followed by the tibia. The leg colors and spines may be useful in identification. The second and third thoracic segments (the mesothroax and metathorax together) are sometimes referred to as the pterothorax (ptero meaning wing) since each segment bears a pair of wings. The abdomen is divided into ten chitinized segments separated by membranous folds, allowing the abdomen to bend and expand. The abdominal segments, from the base to the tip, are referred to as S1 thorough S10, and we use this convention in the figures and charts. The secondary sexual apparatus of the male lies on the bottom of the second abdominal segment (S2). On this complex structure are found the paired hamules which are used in making positive identification of some species in this guide.

The wings are illustrated in Figure 2B showing the important wing venation that is used in the guide. The presence of single or double cells between the radial planate and radial sector above it are used in identification of the whitefaces. The triangles and anal loop are important family characters although you will quickly learn to identify families without them. The shape and color of the pterostigma is also useful for some groups.

Stripes on the top and sides of the thorax (see Figure 2C) can be useful in identification of some species. On the top, a central median stripe is referred to as the mid-dorsal stripe. In some species a top stripe on the anterior of the thorax is referred to as a "front" stripe, as in the Chalk-fronted Corporal. The stripe on the top edge of the thorax is referred to as the shoulder stripe. Stripe(s) in this region may be antehumeral or humeral depending on which side they fall of a major suture in the thorax where two plates come together. The sides of the thorax may be without stripes or may contain one or more stripes referred to as side stripes, technically called lateral stripe(s).

Figures 2D and 2E illustrate the top and front of the head showing the compound eyes, small antennae, and face. The facial line running horizontally across the face and the shape of the "T-spot" on top of the head are important in identification of darners. The occipital horns are useful for identification of some of the snaketail species. In addition to the compound eyes, there are three single eyes (ocelli) on top of the head. The mandibles on the bottom of the head are at least partly covered by chitinized flaps known as the upper and lower lips (labrum and labium). The colored eyespots and occipital bar lying behind the eyes of some of the damsels are useful in aiding in identification.

The terminal appendages of the male are shown in Figure 2F. The upper appendages are modified cerci used to grasp the female during copulation. The size and shape of the cerci are important characters used in identification. The lower appendage in the dragonflies is the epiproct. Damselflies have a pair of lower appendages called paraprocts which are used in identification of spreadwings.

Figure 2G illustrates the anatomy of some of the female Odonata. Females of species that oviposit directly into water often have an enlarged subgenital plate while those that oviposit into plant tissue, woody material, sand, or mud have an ovipositor. The shape of the subgenital plate is important in identification of many species, as is the size, shape, and length of the ovipositor.

Olive Clubtail male

Figure 2 • Odonate Morphology

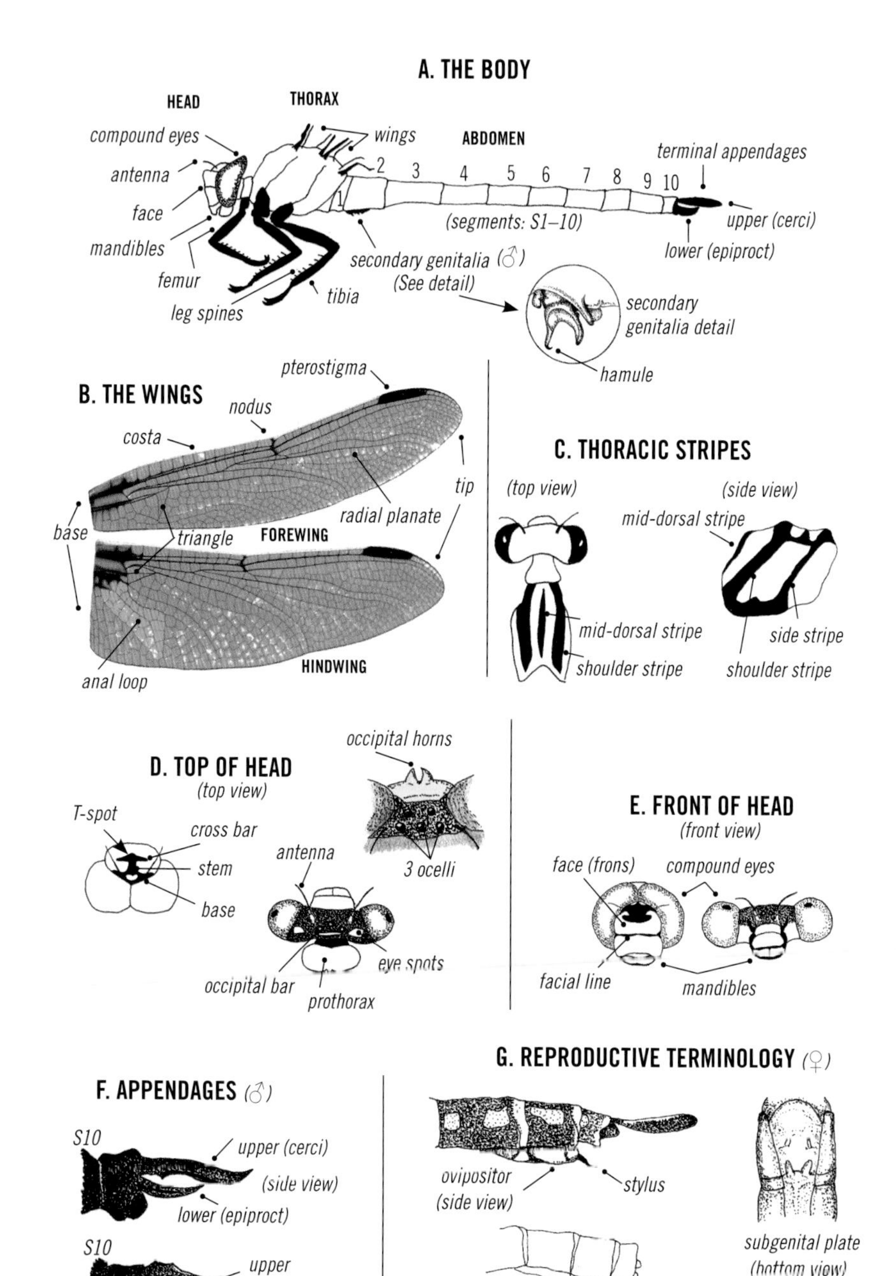

Suborder and Family Keys

The chart on this page illustrates characters used to distinguish the dragonfly and damselfly suborders. A simple annotated key to the families follows.

Identification Chart A • Telling Them Apart

Dragonflies, Suborder Anisoptera, vs. Damselflies, Suborder Zygoptera

Dragonflies (Anisoptera)	Damselflies (Zygoptera)
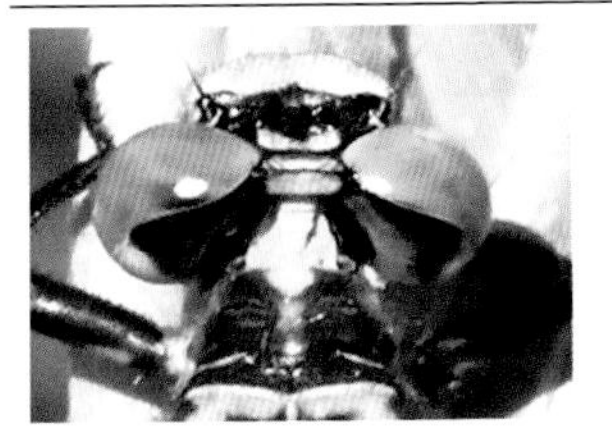	
Compound eyes touch or close together (less than one eye width separation) See page 59.	Compound eyes widely separated. See page 221.
	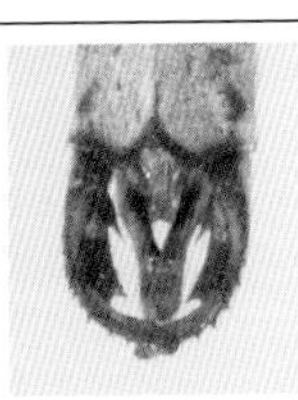
Three abdominal appendages: one epiproct between two cerci (top view, not always visible)	Four abdominal appendages: two paraprocts between two cerci (top view, not always visible)
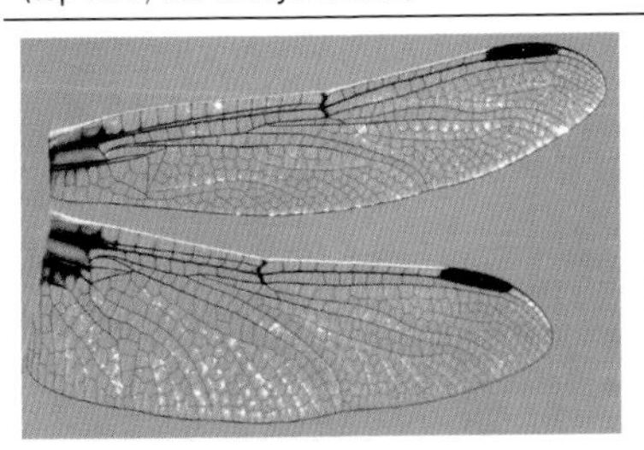	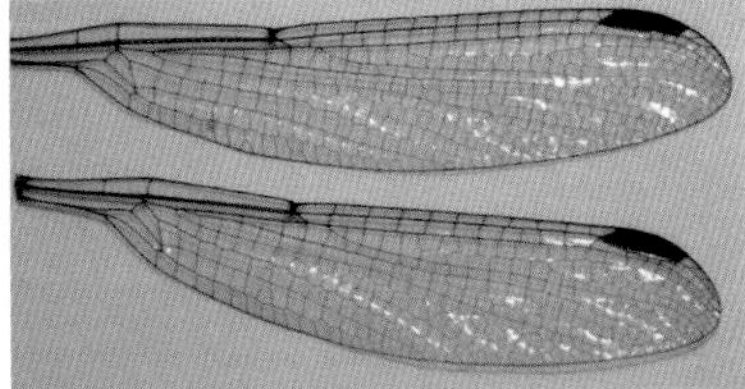
Wings shaped differently	Wings shaped similarly
Wings held horizontal or slightly forward at rest	Wings held along, over, or angled back above abdomen
Robust, strong fliers	Thin, weak fliers
Thick abdomen	Thin abdomen

Identification Chart B • Key to Dragonfly (Anisoptera) Families

Simple Key to Oregon Dragonflies (Anisoptera)

Eyes separated

Long, narrow pterostigma

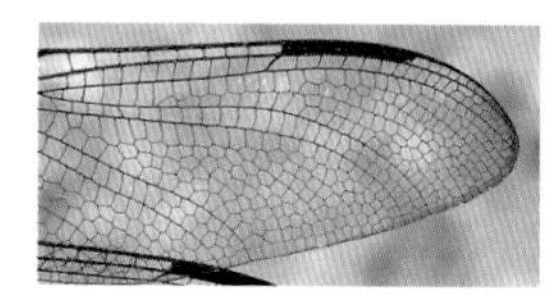

Petaltail (Petaluridae), one species. See page 59.

Eyes well separated

Pterostigma shorter

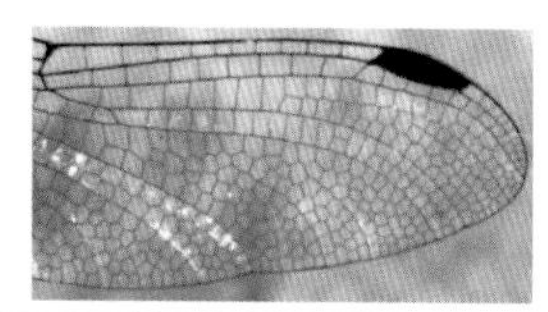

Clubtails, RIngtail, Snaketails (Gomphidae), nine species. See page 104.

Eyes broadly touching

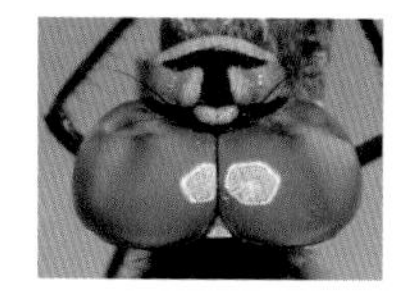

Triangle in front wing long side-to-side

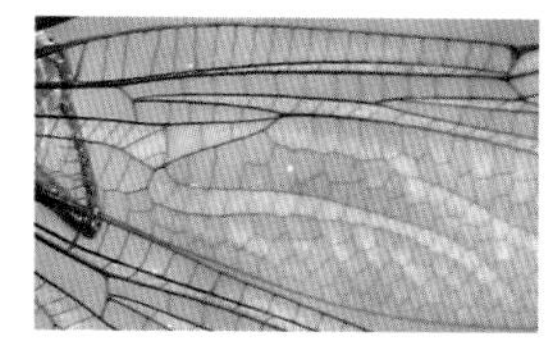

Darners (Aeshnidae), thirteen species. See page 72.

Eyes broadly touching

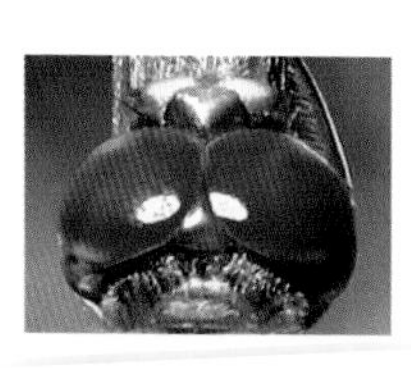

Triangle in front wing long front-to-back

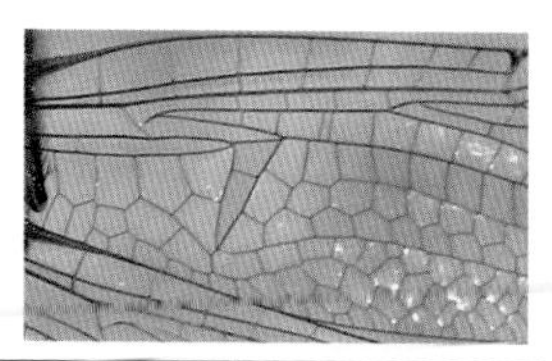

Skimmers (Libellulidae), thirty-one species. See page 147.

Eyes narrowly separated

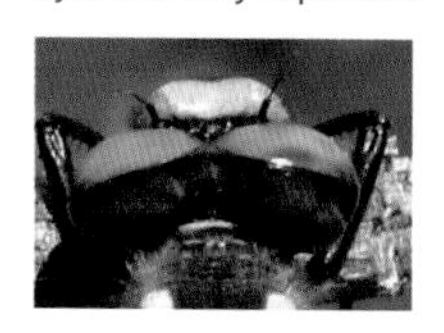

Unique thoracic stripes

Spiketail (Cordulegastridae), one species. See page 65.

(continued on next page)

Identification Chart B • Key to Dragonfly (Anisoptera) Families

(continued from previous page)

Eyes just touching

Anal loop rounded (hindwing)

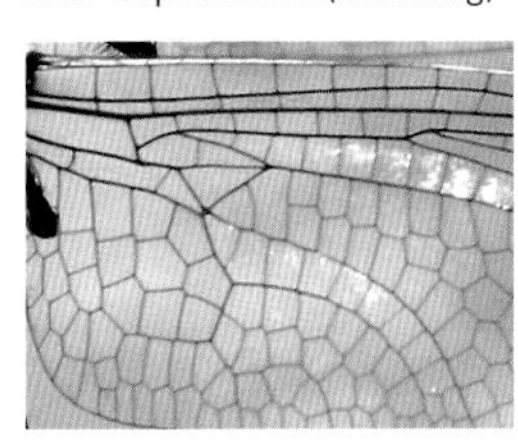

Cruiser (Macromiidae), one species. See page 69.

Eyes just touching

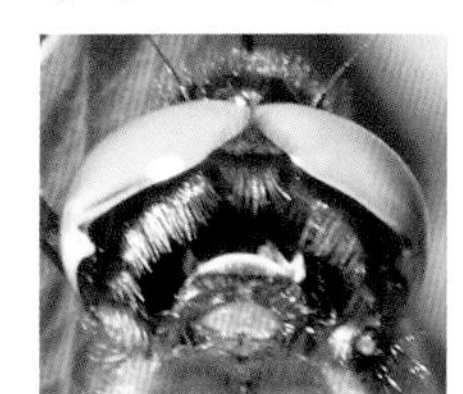

Anal loop elongated (hindwing)

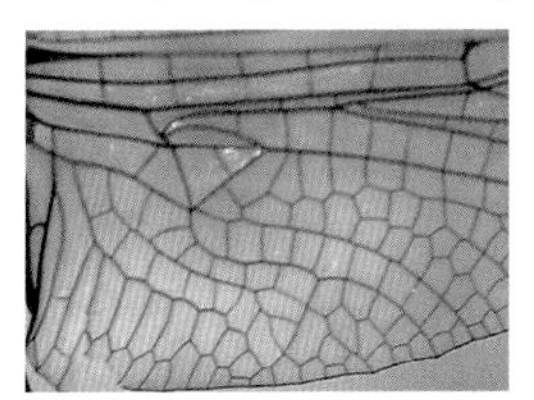

Emeralds (Corduliidae), seven species. See page 128.

Four-spotted Skimmer female

Identification Chart C • Key to Damselfly (Zygoptera) Families

This annotated key to the families is used by matching characters of observed Odonates to those listed in the first column of the chart. If the features match, check the second column to ensure positive identification. In cases where two families share characters in the first column, check the second column for traits to separate the families. The final column lists the family name and number of Oregon species.

Simple Key to Oregon Damselflies (Zygoptera)

Wings broad at base	Wing patterned in black or red	Broad-winged damselflies (Calopterygidae), two species. See page 221.
Wings narrowed at base and wing venation as shown 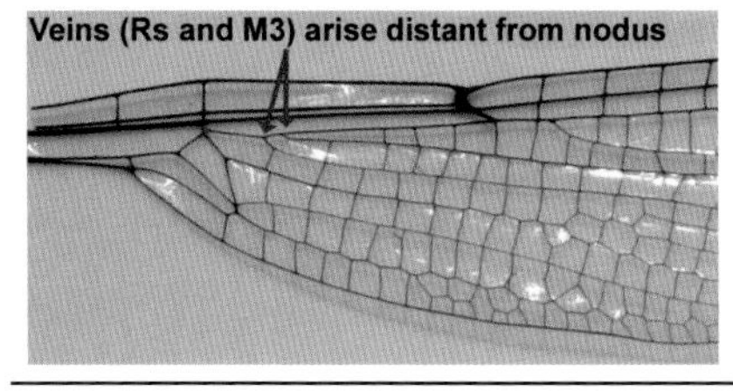 	Wings held open and back when perched	Spreadwings (Lestidae), seven species. See page 226.
Wings narrowed at base and wing venation as shown 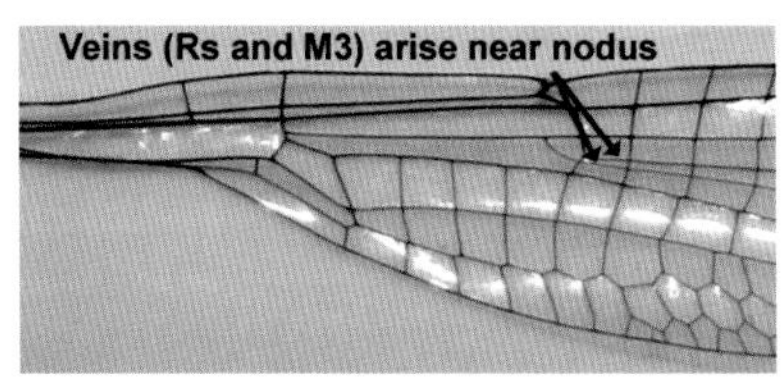 	Wings held closed or over or next to the abdomen when perched	Pond Damsels (Coenagrionidae), nineteen species. See page 242.

II. Oregon Geography and Odonate Distribution

For a better understanding of Oregon dragonfly and damselfly distribution, this section is an introduction to the varied nature of Oregon and some of the complexities of Odonata distribution. Each of Oregon's eight ecoregions is described, and we suggest one or more good places to find dragonflies in each ecoregion with reference to those sites in another section in this field guide, "Oregon's Great Dragonflying Spots."

Oregon is a large, varied, and beautiful state of over 98,000 square miles (61.9 million acres) and is home to almost 3.8 million people. Interstate-5 (I-5) traverses western Oregon from Portland in the north to California on the south, covering a distance of about 310 freeway miles. From Florence on the west side of the state to Ontario on the east side, it is 360 miles as the Spot-winged Glider flies. The Pacific Ocean forms Oregon's western boundary and greatly influences the state's climate. Oregon shares several mountain ranges, drainage basins, and ecoregions with its adjacent states, California, Nevada, Idaho, and Washington.

More than 56 percent of Oregon's land area is publicly owned, the majority by the federal government: the U.S. Bureau of Land Management (BLM) with 15.8 million acres (25.4%), and the U.S. Forest Service (USFS) with 15.6 million acres (25.2%). Other federal, state, and local governments own the remaining 3.4 million acres of public lands, and about 27.2 million acres (43.9%) is privately owned. There are eleven National Forests in Oregon concentrated in the forested, mountainous regions. BLM owns forest lands in western Oregon and much of the land in the Basin and Range region in Oregon's southeastern quarter.

Oregon is divided into two distinct parts by the Cascade Mountain range, and this barrier extends through Oregon from Washington into northern California. The western third of the state is wetter, with a mild climate in comparison with the eastern two-thirds, which is drier and has colder winters and relatively hotter summers. This division has a bearing on dragonfly and damselfly distribution. At least ten species of dragonflies and five species of damselflies are restricted to eastern Oregon:

Dragonflies: Lance-tipped, Subarctic, and Black-tipped Darners; White-belted Ringtail; Columbia Clubtail; Ocellated Emerald; Bleached, Comanche, and Red Rock Skimmers; and Desert Whitetail.

Damselflies: Sweetflag Spreadwing; Aztec and Paiute Dancers; and Alkali and River Bluets.

Only one species of dragonfly is restricted solely to western Oregon, the Bison Snaketail.

Storms blow across western Oregon from late fall into early spring, bringing heavy rainfall at lower elevations and snow at the higher elevations. By the time these storms pass over the Coast Range and Cascade Mountains, less moisture remains and eastern Oregon falls into the rain shadow of the Cascades. The eastern Oregon Mountains and high ridges receive winter snow. Most of eastern Oregon lies at higher elevations (between 2,000 and 4,500 feet) than the western valleys (75 to 1,900 feet), so winter and summer temperatures are more extreme. More continental storms reach eastern Oregon from the north, bringing colder winters and less precipitation. Summers in western Oregon are Mediterranean: dry and warm, with southwestern Oregon being even drier and warmer than northwestern Oregon. The Klamath Mountain region's climate resembles that of northern California.

Oregon is divided into thirty-six counties (see map below). Steve Valley began recording county records in 1990 as an aid to understanding Odonate distribution.

Reports of new county records are now kept by Odonata Central and readers can access maps and individual county lists to learn more about distribution by county (http://odonatacentral.org). Jim Johnson also maintains a current Oregon county list as well as early and late flight dates at this website (http://www.odonata.bogfoot.net/)

While there are ten ecoregions, Oregon can be divided into eight ecoregions for purposes of Odonate discussion (see map on page 29): four in the west and four in the east. However, there are 21 species of dragonflies and 13 species of damselflies that are widespread in Oregon. That is, they are distributed broadly throughout Oregon at proper aquatic habitats and within certain elevation constraints (see map on page 30), and have been recorded in 25 or more of the 36 counties.

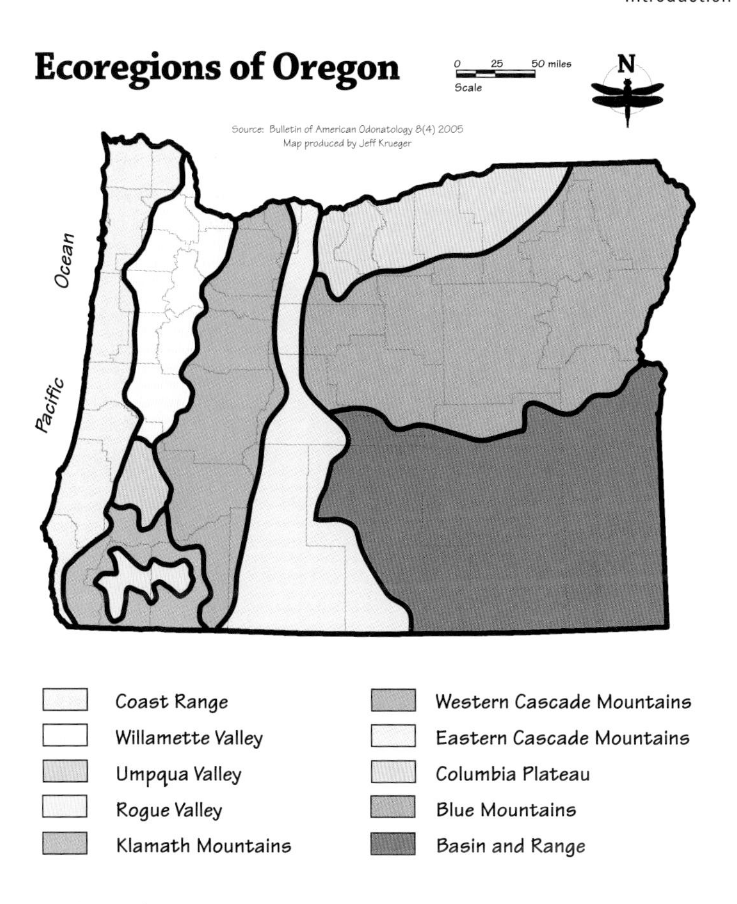

The 34 widespread species are:

Variable Darner	most common in mountains at 3,000'–7,500'
Paddle-tailed Darner	up to 7,500'
Shadow Darner	up to 6,000'
Common Green Darner	common up to 5,000'; uncommon up to 6,000'
California Darner	up to 7,300'
Blue-eyed Darner	common up to 4,000'; uncommon up to 6,000'
Western Pondhawk	below 1,000' in west; up to 4,900' in Basin-Range
Dot-tailed Whiteface	up to 7,200'
Eight-spotted Skimmer	primarily below 1,000'; up to 6,000' in southeast
Twelve-spotted Skimmer	primarily below 1,500'; up to 7,000' in So. Cascades and Basin-Range
Four-spotted Skimmer	more common at higher elevations to 7,000'

Elevation (feet above sea level)

Blue Dasher	below 1,000' in Western OR and in Columbia Plateau
Common Whitetail	common below 1,000' in west; up to 6,000' in So. Cascades and Basin-Range
Variegated Meadowhawk	up to 7,500'
Saffron-winged Meadowhawk	up to 6,100'
Red-veined Meadowhawk	up to 4,500'; common at some coastal dune wetlands
Band-winged Meadowhawk	up to 7,500', Eastern OR and Western Interior Valleys
Striped Meadowhawk	up to 7,500'
Black Saddlebags	up to 5,000'
Wandering Glider	wide distribution but erratic
Spot-winged Glider	wide distribution but erratic
River Jewelwing	north up to 1,000'; east up to 4,600'
California Spreadwing	up to 5,400'; Willamette Valley below 500'
Spotted Spreadwing	up to 5,200'
Northern Spreadwing	up to 7,500'

Emerald Spreadwing	up to 7,500'
Lyre-tipped Spreadwing	up to 7,500'
Western Red Damsel	up to 7,500'
Vivid Dancer	up to 6,300' except NW OR and Coast
Boreal Bluet	400' - 7,500'
Tule Bluet	common up to 1,000'; uncommon up to 7,500'
Northern Bluet	up to 7,500'
Pacific Forktail	up to 7,300'
Western Forktail	up to 7,300'

Another twelve species have very restricted ranges; they have been found at only one to four locations within a narrow range of habitats and ecoregions. Those species are:

Lance-tipped Darner	only at Lake of the Woods, Klamath Co.
Zigzag Darner	3,000'–5,000' in four counties.
Subarctic Darner	3,000'–3,500' in Clackamas Co.
Black-tipped Darner	only from a borrow pit, Union Co.
Ocellated Emerald	4,600'–6,000'; Deschutes and Klamath Cos.
Brush-tipped Emerald	3,200'–4,900' in the Cascade Mountains
Belted Whiteface	only at Magone Lake, Grant Co.
Comanche Skimmer	only from two hot springs, Harney Co.
Bleached Skimmer	only from three hot springs, Harney Co.
Red Rock Skimmer	only from Cottonwood Creek, Harney Co.
Sweetflag Spreadwing	only from Union and Wallowa Cos.
Aztec Dancer	only from one hot spring at Twentymile Creek, Lake Co.

The remaining half (45 species) of Oregon's Odonates have a more complex distribution pattern. Refer to the individual species descriptions and range maps for more information.

Coast Range. This region encompasses the narrow coastal plain and the relatively low Coast Range Mountains that extend from the Columbia River south to the Klamath Mountains near Gold Beach. Comprised of marine sediments and oceanic volcanics, these mountains are still being uplifted as the Pacific Tectonic plate and the San Juan de Fuca microplate push under the North American plate. The narrow coastal plain includes habitat for Odonates with its estuaries, interdunal wetlands, and small coastal lakes. Within the fog-shrouded areas along the coast, including rocky headlands, are forests dominated by Sitka spruce. In southernmost Oregon, along the Winchuck River, is a small redwood zone. The Coast Range Mountains are vegetated by rain-soaked coniferous forests of Douglas-fir, western hemlock, and western red cedar with red alder and bigleaf maple in the understory. Coastal communities receive 60 to 80 inches of rain per year. As the Pacific storms push saturated air masses over the Coast Range, the orographic effect results in rainfall of 180-200 inches annually above the 2,000 foot level. Two of the highest points in the Coast Range are Saddle Mountain (3,203 feet) and Marys Peak (4,097 feet). Many of Oregon's great rivers cut through the Coast Range to flow into the Pacific Ocean: the Columbia, Siuslaw, Umpqua, and Rogue. Other coastal rivers flow from the Coast Range directly to the sea: Necanicum, Young, Nehalem, Nestucca, Trask, Wilson, Siletz, Yaquina, Alsea, Coos, Coquille, Sixes, and Elk.

Red-veined Meadowhawks are common in the wetlands among the dunes. Otherwise, relatively cool summers and lack of lakes in the Coast Range mountains limit this region to the more widespread species. A recent discovery of Black Petaltail should lead to more exploration of suitable Coast Range habitats. Also, the fall migration of dragonflies southward along the coast is not well understood and deserves more study.

See the "Oregon's Great Dragonflying Spots" section under Garrison Lake and Arizona Beach in Curry County, Ten Mile Lake in Coos County, and Sunset Beach in Clatsop County for good coastal spots to find Odonates.

Western Interior Valleys. The Willamette Valley ecoregion, the lowland portions of the Umpqua River Basin near Roseburg, and the Rogue River Basin near Medford and Grants Pass are lumped into this section because of similarity in shared Odonate species and the ability of species to expand northward through these lowland areas. These three valley regions are similar in that they fall within the rain shadow of either the Coast Range or Klamath Mountains on the west, and the western Cascades lie to the east of each of them. Most locations in the Willamette Valley receive between 35 and 45 inches of annual precipitation. Roseburg receives 32 inches, Grants pass 31 inches, and Medford 19 inches annually. The Willamette River flows north past Eugene (422 feet elevation) and Portland (72 feet elevation) before joining the Columbia River. Roseburg is at 465 feet elevation, Grants Pass at 960 feet, Medford at 1,382 feet, and Ashland at 1,895 feet. All three valleys support agriculture, with the Willamette Valley being the most productive agricultural region in Oregon. The Willamette Valley is underlain with non-marine sediments and flood (flowing) lava. Roseburg is situated on marine sediments and Medford is located on both marine and non-marine sediments. Medford is famous for its orchard crops, but all three valleys now support a growing wine industry. Western Oregon is also more populous with 3.2 million (86.7 percent) of Oregonians living there; the major urban areas are concentrated along the I-5 corridor. Over 70% of Oregon's population resides in the Willamette Valley. Vegetation and water availability change as you move south along the I-5 corridor. The Willamette Valley is dominated by Douglas-fir on the foothills. In Roseburg and Medford, the vegetation is like that described in the Klamath Mountains section, but much of the Umpqua and Rogue Valleys are devoted to agriculture. Water features are more plentiful in the Willamette Valley with all its tributary rivers and creeks: Coast and Middle Fork Willamette, McKenzie, Muddy, Marys, Calapuya, Santiam, Clackamas, Yamhill, Tualatin, Sandy, Molalla, and Pudding. Hydric soils are abundant on the fringes of the Willamette Valley—once a region of mixed upland and wet prairies and open oak savannah habitats.

Fifty species of Odonates have been recorded in the Willamette Valley. During the past twenty years both Widow and Flame Skimmers have experienced northward range expansion and can be found now throughout the Valley. In

September and October, Autumn Meadowhawk can be abundant. In the southern Willamette Valley, low elevation streams harbor American Rubyspot, as well as Gomphids: Grappletail, Sinuous and Pale Snaketails in addition to Pacific Clubtail. Both Pacific Spiketail and Western River Cruiser can be found in the Willamette Valley. Along the Columbia River, Olive Clubtail can be found, and there is a recent photographic record of Olive Clubtail from Linn County. The Umpqua and Rogue River Valleys are the best places to find Bison Snaketail, American Rubyspot, Black Spreadwing, California and Sooty Dancers, and Familiar Bluet.

See the "Oregon's Great Dragonflying Spots" section for a number of Interior Valley places to find Odonates: Freeway Lakes in Linn County; Sandpiper Pond and Mt. Pisgah in Lane County; the South Umpqua River at Myrtle Creek in Douglas County; Medford Sports Park in Jackson County; Illinois River Forks State Park in Josephine County; and a list of good spots in the Portland Metropolitan Region.

Klamath Mountains. Oregon's portion of the Klamath Mountains represents 40% of the larger 12,000-square-mile region which extends into northwestern California. Among the oldest mountains in Oregon, this region extends northward from the California border to north of Roseburg and from the southwestern coast near Brookings eastward to the western Cascades at Bear Creek near Medford. The Siskiyou Mountains comprise the southern portion of the range in Oregon. Mt. Ashland at 7,530 feet elevation is the highest point. The rugged Klamaths include igneous, sedimentary, and some metamorphic rocks (like green serpentine) that have been folded and deeply eroded. The main drainage basins in the region are the Umpqua, Rogue, Illinois, and Applegate Rivers. The Chetco River drains westward to the Pacific Ocean at Brookings. Summers in this region are hot and dry, almost drought-like. Occasional severe winter storms can close Siskiyou Pass on I-5 (elevation 4,310 feet). The vegetation is varied with Douglas-fir; incense and western red cedar; ponderosa, Jeffrey, sugar, knobcone, and western white pine; Oregon white oak; Pacific madrone and bigleaf maple. This region is so rugged that the Siskiyou National Forest brochures warn of the dangers of thirst, black bears, cougars, yellowjackets, rattlesnakes, and poison oak—and not necessarily in that order!

Of major interest in the Klamath/Siskiyou Mountains are *Darlingtonia californica* (pitcher plant) fens where Black Petaltails can be found. In this ecoregion, also look for Chalk-fronted Corporal and Hoary Skimmer.

In the "Oregon's Great Dragonflying Spots" section, see the Illinois River at Eight Dollar Mountain and the Darlingtonia Fens in Josephine County for good Odonate sites in the Klamath ecoregion.

Western Cascades. This region extends from Washington, through the entire north-south length of Oregon, and into northern California. This spectacular mountain range is comprised of volcanics with dramatic peaks including Oregon's highest, Mt. Hood at 11,145 feet. Many other volcanic peaks rise above 7,000 feet in elevation. Among the tallest are Mt. Jefferson (10,497 feet), the Three Sisters (North 10,085, Middle 10,497, and South 10,358 feet respectively), and Mt. McLoughlin at 9,495 feet. Crater Lake National Park at 6,176 feet is the site of ancient Mt. Mazama and home to America's deepest lake at 1,943 feet deep. Many of the mountain passes are nearly one mile in elevation: Santiam (Highway 20) at 4,817 feet, Willamette (Highway 58) at 5,128 feet, and Cascade Summit (Highway 138) at 5,920 feet. Vegetation at the temperate lower elevations is Douglas-fir/Oregon white oak transitioning to Douglas-fir/western hemlock. Near the crest are subalpine forests and meadows with alpine conditions at the highest elevations. During the last Ice Age, glaciers covered the Cascades. Annual precipitation is over 80 inches in much of the range, with substantial winter snow pack at higher elevations. Crater Lake National Park averages just over 44 feet of snow annually. The region is dotted with lakes and is the source of numerous rivers: Clackamas, Molalla, Santiam, McKenzie, Willamette, Umpqua, and Rogue.

The western Cascades region is a good place to find Taiga Bluet; Sedge Darner; Grappeltail; American, Ringed, and Mountain Emeralds; Pacific Spiketail; Hudsonian and Crimson-ringed Whitefaces; and White-faced and Black Meadowhawks. Also look for Swift Forktail at up to 4,800-foot elevations. In Jackson County at Jenny Creek, along Highway 66, look for Walker's Darner.

A good spot for this ecoregion referenced in the "Oregon's Great Dragonflying Spots" section is Gold Lake in Lane County.

Eastern Cascades. This region extends from Washington to California along the eastern crest of the Cascade Range. Like the western Cascades, it is of volcanic origin with basalt flows extending eastward from the crest into the High Lava Plains. The snow-covered and glaciated volcanic Cascade peaks are even more beautiful and dramatic when viewed from the east side; and the Cascade Lakes Highway out of Bend is one of the most scenic drives in Oregon. The region is dotted with many volcanic cones and buttes. Sitting in the rain shadow of the Cascades, this region is drier than the west side. Bend, Klamath Falls and Lakeview receive 12 to 16 inches of annual precipitation. Bend lies at 3,628 feet elevation, Klamath Falls at 4,120 feet, and Lakeview at 4,802 feet. The wonderful Deschutes and Hood Rivers flow through the northern section of this region. Drainage in the southern section is more complex; the Williamson and Sprague Rivers flow into the Upper Klamath Lake, and the Lost River, once part of the Great Basin, is now diverted into the Klamath system. The northern portion of large Goose Lake along the California border actually drains southward into California's upper Sacramento River. Vegetation includes conifer/oak forests near Hood River in the north. One of the treats in this region is the open ponderosa pine forests, and on the well-drained pumice plateaus, lodgepole pine is the dominant forest cover. On the eastern edges of the region are areas dominated by juniper/sagebrush and bitterbrush.

A number of species have been found only in the Cascades. Many are at higher altitudes and may slip over the crest in either the Western or Eastern Cascade ecoregion where habitat is suitable. Among those species are: Canada, Zigzag, and Subarctic Darners; Ocellated and Brush-tipped Emeralds; and Sedge Sprite. Several interesting species have been found in the southern portion of the eastern Cascades ecoregion: Lance-tipped and Walker's Darners; Pacific Clubtail, Grappletail; and Hoary Skimmer.

There are several scenic sites that are also good for Odonates mentioned in the "Oregon's Great Dragonflying Spots" section: Camas Prairie in Wasco County; Todd Lake and the Cascade Lakes Highway in Deschutes County;

Jenny Creek in Jackson County, and Crescent Creek and Lake of the Woods in Klamath County.

Columbia Plateau. Located in north-central Oregon south of the Columbia River, this region extends eastward from The Dalles to just beyond Pendleton with the Blue Mountains to the south. It is an even-crested plateau ranging from 1,200 to 1,800 feet high with the steep, deeply carved canyons of the lower Deschutes and John Day Rivers in its western portion and Willow Creek and the Umatilla River in its eastern portion. These drainages flow from higher elevations in the eastern Cascades and Blue Mountain regions. The lowest elevations in this region are along the Columbia River with Umatilla at 296 feet and The Dalles at 92 feet above sea level. This plateau is underlain by thick lava rock and portions are covered by wind-blown loess deposits. This arid region receives from 9 to 14 inches of annual precipitation. Originally vegetated by sagebrush steppe and grasslands, the Columbia Plateau is almost entirely privately owned and devoted to agricultural production with wheat being the primary crop. The entire region is almost devoid of forest. Some public access is available at BLM lands along the lower Deschutes and John Day Rivers and along the Columbia River, at three National Wildlife Refuges near Umatilla and Pendleton, and in a few State Parks. Other than the above mentioned rivers, parks, and refuges, access to water habitat is scarce in this region.

Species to look for in this region include: Columbia and Olive Clubtails; Sinuous Snaketail; White-belted Ringtail; Blue Dasher; Cardinal Meadowhawk; and Widow Skimmer. A few specimens of Walker's Darner have been found near the Columbia River in Gilliam and Sherman Counties.

While any of the public lands along the Columbia River are worth checking, the John Day River at Cottonwood in Gilliam County is detailed in the "Oregon's Great Dragonflying Spots" section.

Blue Mountains. This region of complex mountain ranges (Ochoco, Strawberry, Blue, Wallowa, and Elkhorn) is of mostly volcanic origin, but in places metamorphic rocks and intrusive granites are exposed (in the Wallowas and Elkhorns). A small portion of the Blue Mountains extends into southeastern Washington. Several major peaks exceed 9,000 feet in elevation, including Eagle Cap at 9,595 feet and Strawberry Mountain at 9,038 feet. During the last Ice Age, the Wallowas experienced significant glaciation. Here cool summers are matched by severe winters. Higher elevations receive 20 to 30 inches of precipitation annually and Eagle Cap gets over 70 inches. Lower elevations are much drier: Fossil gets 15 inches and Baker City only 11 inches annually. A number of drainages originate in the Blue Mountains: Burnt, Powder, John Day, Umatilla, Grande Ronde, Imnaha, Crooked, and Silvies Rivers. At 281 miles, the John Day is Oregon's longest river wholly contained within the state—and the third-longest undammed river in the contiguous U.S. Some of these rivers drain north to the Columbia, some east to the Snake, and others south into the Basin and Range Region. Vegetation is varied with ponderosa pine, true firs, western larch, juniper, sagebrush, bitterbrush, bunchgrass, subalpine forests, and alpine habitats at the highest altitudes. A July drive along USFS Road No. 42 out of Ochoco is so full of roadside yellow daisies, purple and blue lupine, red paintbrush and yellow, orange and white butterflies that it could make an Odonatist want to become a Botanist or Lepidopterist. One of the spectacular wonders in the northeast corner of this region is Hells Canyon of the Snake River, shared with Idaho. It is America's deepest gorge at over 8,000 feet. In one 50-mile stretch its depth averages 5,500 feet.

Species of interest in this region include: Sedge and Black-tipped Darners; Belted, Hudsonian, and Crimson-waisted Whitefaces; Ringed and Mountain Emeralds; White-faced, Cherry-faced, and Autumn Meadowhawks; Sweetflag Spreadwing; and Taiga Bluet.

This varied region has a number of sites in the "Oregon's Greatest Dragonflying Spots" section: Anthony Lakes in Baker County; Grand Ronde Lake, Mud Lake, and Floodplain Flats in Union County; Magone Lake in Grant County; the John Day River at Muleshoe in Wheeler County; Borrow Pit in Union County; and Enterprise Wetland, Wallowa County.

Basin and Range. This vast region covers the southeastern corner of Oregon and includes the subregions of High Lava Plains and Owyhee Uplands. Much of the area is owned by the Bureau of Land Management with inholdings of private lands, used primarily for cattle grazing and pasture crops. More intense agriculture can be found near Ontario where irrigation sources are available. Outside of Ontario, Vale, and Burns, the entire region is sparsely populated. This is a dry landscape of sagebrush, juniper, and rimrock lying at about 4,000 feet above sea level. At higher elevations some forests exist, with beautiful stands of quaking aspen on Steens Mountain, which rises to its highest point at 9,773 feet above sea level. The geology is of volcanic origin with vast lava flows covering much of the region. Large northwest-southeast trending fault blocks give the region its name. Below the steep sides of these fault blocks lie remnant lakes (e.g., Summer, Abert, and the Alvord, which is mostly a dry playa except in the wettest years). During the Pleistocene Era, these lakes were vast and more plentiful. Summer and Abert Lakes were once part of the larger Chewaucan. Most famous of the marshes in this region are those along the Donner und Blitzen River at Malheur National Wildlife Refuge at the south end of Malheur Lake. Burns receives about 12.5 inches of annual precipitation and both Vale and Ontario (at 2,140 feet above sea level) receive less than 10 inches. The driest place in Oregon is the Alvord Basin which averages 5 inches of rain per year; Alvord Lake is at 4,020 feet elevation. It is here, at the base of the Steens and Pueblo Mountains, that rocky creeks and hot springs provide unique Oregon Odonate habitats. The southeastern corner of this Basin and Range region occupies the northern 5 percent of the Great Basin. This is an area with internal drainage with almost no outlet to the Pacific Ocean. Among the rivers and major creeks in the region are the Silvies, Donner and Blitzen, Owyhee, Silver, and Malheur—one of the few that does drain to the Pacific Ocean, via the Snake and Columbia Rivers.

Several southwestern U.S. species reach the northern extent of their ranges in southeastern Oregon. This makes the array of species in the Basin and Range ecoregion especially interesting: Comanche, Bleached, Red Rock, and Hoary Skimmers; Desert Whitetail; Aztec and Paiute Dancers; Alkali and River Bluets; and Black-fronted Forktail are all regular or have been recorded. Several of these species have very restricted distribution in Oregon, often limited to alkaline lakes and desert hot spring habitats. The Owyhee River basin is a good place to find White-belted Ringtail, Columbia Clubtail, Sinuous Snaketail, and Western River

Cruiser. Pacific Spiketail has been found along creeks flowing from the Pueblo and Steens Mountains— one of the few Basin and Range locations for this species.

Four locations from this ecoregion are covered in the "Oregon's Great Dragonflying Spots" section: Borax Lake and Mickey Hot Spring in Harney County; Owyhee River at Rome in Malheur County; and Twentymile Creek in Lake County.

Getting Closer to Oregon's Odonates

What's in a name?

Oregon dragonfly survey in the Ochoco Mountains, 2007. From left: Steve Berliner, Josh Vlach, Chris Marshall, Steve Valley, Steve Gordon, Jim Johnson, and Cary Kerst • photo by Steve Valley

From the time children begin talking, they start to name things. We have an innate desire to learn the names of things and places in our world. Names are also very important to scientists as they allow everyone to have certainty about the exact organism that is being researched or discussed. For this reason, the system of scientific nomenclature was developed. Each organism is named according to strict international protocols with a genus and species name.

The scientific names, most often of Greek or Latin origin, are cumbersome for some people. Dennis Paulson and Syd Dunkle began developing a list of common or English names for the Odonata in 1978. This list was eventually adopted by the Dragonfly Society of the Americas (DSA) in August 1996 at its annual meeting. The list has been maintained and updated as necessary through the years by the DSA. This allows for the more precise use of common or English names, and has aided the popularity of dragonfly watching in the United States. We will use these English names in this book but will also provide scientific names for reference for those who wish to learn both.

Dragonfly Watching

The interest in watching dragonflies and damselflies (aka dragonflying) has increased dramatically in recent years. It has become a recreational activity similar to birdwatching (aka birding) with an interest in seeing new species and keeping lists of observed species. It is not necessary to travel great distances—many species can be found in your local area near ponds, streams, and wetlands, or even in the backyard. Watching dragonflies is a great way to relieve a stressful day.

A good pair of close-focusing binoculars is a great asset for identifying and observing these interesting insects and can be used for bird watching too. Remember when approaching Odonates that they have evolved to be keen predators and, thus, are able to see motion very well. It is best to approach them very slowly. The eye of the dragonfly wraps around the head so they are able to see to the side and even to their back to some extent. There is a small area to the rear where there is a blind spot. One can often approach from the rear with some success.

Should you wish to keep a list of species observed, checklists of species by county and state can be found on the Odonata Central website (www.odonatacentral.org). An Oregon checklist is also provided at the back of this book (Appendix A, pages 289–291).

Jim Johnson photographing Sweetflag Spreadwing at its discovery site near Enterprise, OR

Photography

If you enjoy photography, dragonflies and damselflies provide a wonderful opportunity to test your skill. With the advent of digital cameras, which require no film processing, today's photographers can shoot a nearly limitless number of images.

Equipment can be as simple or complex as one wants. Most serious Odonata photographers are using digital SLR cameras with a 180 mm macro lens, a 1.4 tele-converter, and a tripod. Some are able to hand-hold cameras with success. We either hand-hold the camera or use a monopod.

Another method that is being used with success is digiscoping. The spotting scopes used by birdwatchers are fitted with an adapter so that a small digital camera can be used. This can produce some spectacular images with great magnification. It can be more difficult to get set up, but photos can be taken at a greater distance.

Collecting

There are also those who may want to collect specimens for a reference collection. This requires some dedication on the part of the collector to follow guidelines for processing specimens. These guidelines can be found at the Dragonfly Society of the Americas website at Odonata Central. In addition to properly processing specimens, collectors should make a commitment to maintaining a collection so that it will be useful to science in the future. There are no dragonfly or damselfly species listed as

Threatened or Endangered in Oregon. The Columbia Clubtail is listed as "a species of concern" due to its limited distribution within three river watersheds in Oregon and Washington.

While collecting specimens is an activity that draws criticism today, we should remember that much of the information that we know about dragonflies and damselflies comes from scientific collections, descriptions and study associated with collections. For many people, collection is neither necessary nor desired. You can thoroughly enjoy watching through binoculars or photographing Odonates without collecting them.

Steve Valley at Gold Lake Bog (note darner on cap) • photo by Steve Berliner

During the right season, you can find dragonflies and damselflies at or near almost any water feature regardless of its type or size. Garden ponds; roadside trickles; intermittent streams; wet prairies, bogs, and marshes; and rivers can all provide habitat for Odonates. A great part of dragonflying is exploring new places and making your own discoveries.

Oregon's Great Dragonflying Spots

We suggest thirty great Oregon places to observe Odonates (see map on page 40). These sites include at least one site from each of Oregon's eight ecoregions in twenty-one counties. The sites are divided almost equally between eastern and western Oregon. These are among the best places in Oregon to find not only the common and abundant species, but also the most unusual ones. If you pay close attention to the species listed in all the individual sites, you will notice mention of all 91 Oregon species.

Visiting these sites will take you to some of Oregon's most spectacular country: the Elkhorn, Wallowa, Blue, and Cascade Mountains; beautiful mountain and coastal lakes; the Willamette, Umpqua, Illinois, John Day, and Owyhee Rivers; unique bogs, marshes, and fens; and hot springs in the Basin and Range region. In the Alvord Desert under the rain shadow of magnificent Steens Mountain, you can see dark rain clouds dropping rain toward the earth. You can wait a long time and the rain will never reach you or the ground—only five inches of it is measured annually. Here horned lizards scurry under the sage, and Bleached, Hoary, and Comanche Skimmers, Paiute Dancers, and Black-fronted Forktails fly around the isolated wet spots.

You can shiver in August during a storm at 7,000 feet at a rugged mountain lake, waiting anxiously for the sun to appear, the wind to calm down, and the dragonflies to become active. Then you can wade across an icy stream, and through a cold, ankle-deep marsh to search for that dazzling darner.

For each site we give directions, a list of likely species, county, ecoregion, GPS location, elevation, and reference to the *DeLorme Oregon Atlas & Gazetteer* page.

Thirty of Oregon's Great Dragonflying Spots

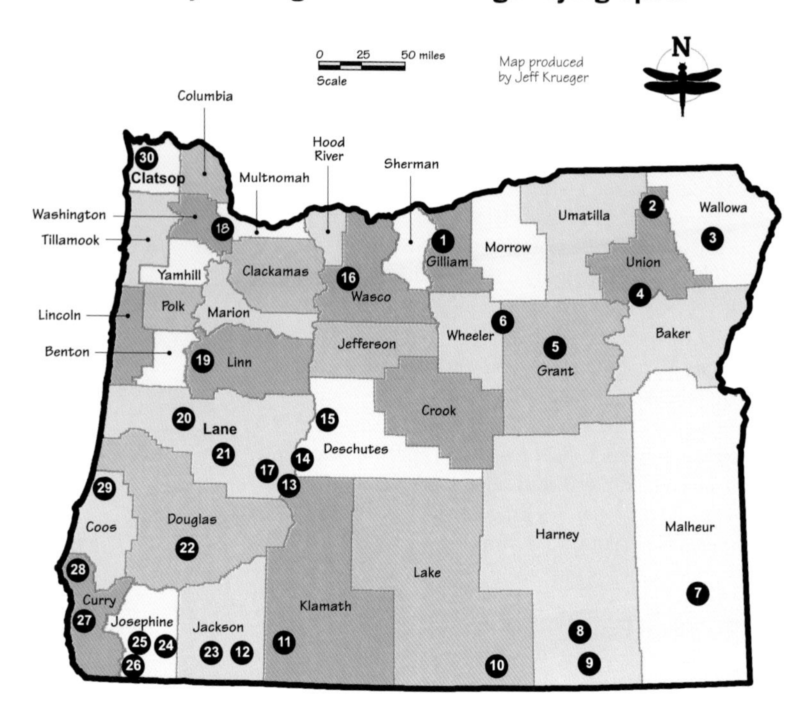

Columbia Plateau Ecoregion

1. John Day River at Cottonwood

Blue Mountains Ecoregion

2. Borrow Pit
3. Wetland, Enterprise
4. Anthony Lake Grande Ronde Lake, Mud Lake, Floodplain Flats
5. Magone Lake
6. John Day River at Muleshoe

Basin and Range Ecoregion

7. Owyhee River at Rome
8. Mickey Hot Springs
9. Borax Lake
10. Twentymile Creek

Eastern Cascade Mountains Ecoregion

11. Lake of the Woods
12. Jenny Creek
13. Crescent Creek
14. Cascade Lakes Highway
15. Todd Lake
16. Camas Prairie

Western Cascade Mountains Ecoregion

17. Gold Lake, Bog, Salt Creek

Interior Valleys (Willamette, Umpqua, Rogue)

18. Portland Metropolitan Region (Oaks Bottom, Beggar's Tick Marsh, Oregon Episcopal School, Hedges Creek, Koll Creekside, Jackson Bottom, Fernhill Wetlands)
    See map below.
19. Freeway Lakes, Albany
20. Sandpiper Pond, Eugene
21. Mt. Pisgah Arboretum
22. South Umpqua River, Myrtle Creek
23. Jackson County Sports Park, Medford
24. Illinois River Forks State Park, Cave Junction

Klamath Mountains (Siskiyous) Ecoregion

25. Illinois River at Eight Dollar Mountain
26. Darlingtonia Fens

Coast Range Ecoregion

27. Arizona Beach State Park
28. Garrison Lake, Port Orford
29. Ten Mile Lake
30. Sunset Beach State Recreation Area

Portland Region Dragonflying Spots

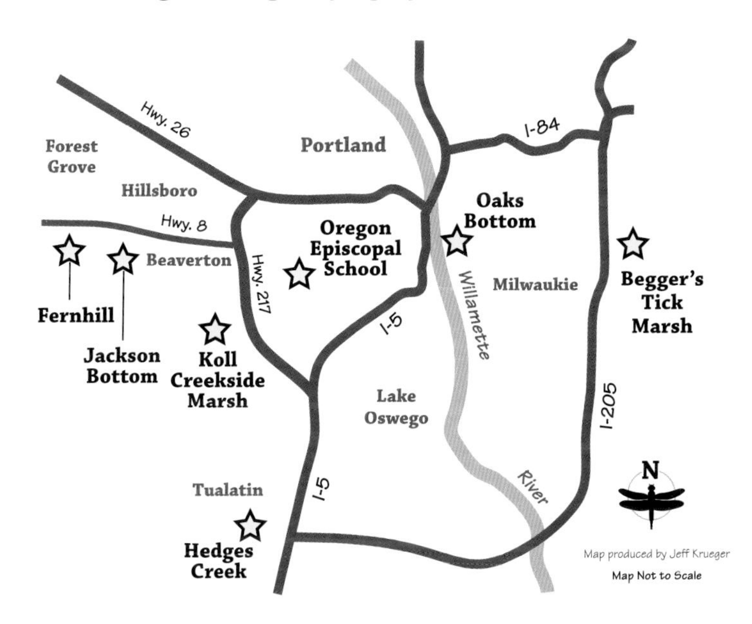

1 John Day River at Cottonwood, Gilliam County *

Highway 206 crosses the John Day River at Cottonwood Canyon 11 miles east of US 97 in Gilliam County. There is good access to the river at this spot with extensive gravel bars during the summer. Early in the year, the river can be bank full. This is also a popular rafting destination. During July and August, this is a great spot to observe White-belted Ringtails, one of our most striking dragonflies. Look for ringtails perched in the sagebrush near the river with the abdomen raised like a flag. It can be very hot in this canyon during the summer, so be prepared.

Other species that may be seen here include Sinuous Snaketail, Eight- and Twelve-spotted Skimmers, Striped Meadowhawk, Emma's Dancer, and Western Forktail, among others.

Ecoregion: Columbia Plateau
Lat: 45.4771°; Long: 120.4677°; Elev: 538 ft.
DeLorme Oregon Atlas & Gazetteer, page 84.

2 Borrow Pit, Union County

This site is somewhat remote and is located approximately 22 miles north of Elgin, Union County along National Forest Road 6231 which is accessed via Lookout Mountain Road. Look for the pond on the east side of the road. In 2009, Jim Johnson discovered the Black-tipped Darner at this site for the first Oregon record for the species. It is a particularly interesting discovery, as the habitat here is a water-filled borrow pit. The small pond has an unusually large population of darners including Black-tipped, Paddle-tailed, and Variable Darners. The Sweetflag Spreadwing can also be seen here for the second known Oregon location for the species. Other species observed here include: Mountain Emerald, Striped Meadowhawk, Northern Spreadwing, Boreal Bluet, and Western Forktail.

Ecoregion: Blue Mountains
Lat: 45.7658°; Long: 117.8211°; Elev: 4,100 ft.
DeLorme Oregon Atlas & Gazetteer, p. 87.

3 Wetland, Enterprise, Wallowa County

Although some distance from population centers, the northeast corner of Oregon offers unique beauty and interesting habitat. The Sweetflag Spreadwing was discovered for the first time in Oregon (2009) at a small wetland 13.6 miles north of Enterprise on the east side of Highway 3. This spreadwing is easy to see at this wetland as there is a large population there. Although small, the site has a diverse population of Odonata, and the list will likely expand with additional visits. Some other species seen here include a number of other spreadwings, such as Emerald, Lyre-tipped, and Northern. Additional species include Paddle-tailed and Variable Darners, Mountain Emerald, Dot-tailed Whiteface, Four-spotted Skimmer, Common Whitetail, Striped Meadowhawk, Boreal Bluet, and Western Forktail.

Ecoregion: Blue Mountains
Lat: 45.6236°; Long: 117.2651°; Elev: 4,328 ft.
DeLorme Oregon Atlas & Gazetteer, p. 87.

---

* *Map sites can be found on page 40–41*

4

Anthony Lake, Baker County
Grande Ronde Lake, Mud Lake, Floodplain Flat, Union County

This group of beautiful, high-elevation lakes lies about 45 miles southeast of LaGrande in the Elkhorn Mountains. These lakes can be reached by taking I-84 south of LaGrande for 25 miles. Take Exit 285 and turn right onto US 30. Turn left onto Ellis Road at about four miles, then right on Anthony Lakes Highway. Follow the signs to Anthony Lake. Mud Lake is just 0.3 miles north of Anthony Lake, and Floodplain Flat is just north of Anthony Lakes Highway prior to reaching the lake. Grand Ronde Lake can be reached via NFD 43 (Ladd Canyon) about a mile past Anthony Lake.

The emeralds can be especially common at these lakes, including American, Mountain, and Ringed. Watch for a number of other species including Paddle-tailed, Sedge, and Variable Darners; Hudsonian Whiteface, Four-spotted Skimmer, Black and Striped Meadowhawks; Emerald, Lyre-tipped, and Northern Spreadwings; Boreal and Taiga Bluets; and Western Forktail. At Floodplain Flats, Swift Forktail may be found in addition to many of the above species.

Ecoregion: Blue Mountains
Anthony Lake: Lat: 44.96141°; Long; 118.23015°; Elev: 7,150 ft.
Grand Ronde Lake: Lat: 44.97735°; Long: 118.24221°; Elev: 7,160 ft.
Mud Lake: Lat: 44.96689°; Long: 118.23345°; Elev: 7,100 ft.
Floodplain Flats: Lat: 44.96424°; Long: 118.22825°; Elev: 7,142 ft.
DeLorme Oregon Atlas & Gazetteer, p. 82.

5

Magone Lake, Grant County

Magone Lake is nestled high in the Blue Mountains about 10 miles north of John Day. This is the only known Oregon location for Belted Whiteface, which was first recorded here in 2008 (Oregon species #89). Named after Major Joseph W. Magone of Canyon City, this lake is pronounced "Magoon" by the locals. To reach the lake on paved roads, travel east from John Day on Highway 26 about 9 miles and turn north onto USFS Road # 18 (Keeny Forks Road) along Bear Creek. After about 4 miles, the paved road veers northwesterly for another 8 miles to the turn off to Magone Lake which is approximately 2 miles further to the southwest. At 4,907 feet in elevation, naturally formed Magone Lake is complete with a boat ramp and campground administered by the Malheur National Forest. A nice trail circles the lake.

In addition to a healthy population of Belted Whiteface, look for Common Green and California Darners; Pacific Spiketail; Dot-tailed Whiteface; Four, Eight, and Twelve-spotted Skimmers; River Jewelwing; Emma's Dancer; Boreal Bluet; and Western Forktail.

Ecoregion: Blue Mountains
Lat: N44.5492°; Long: W118.9132°; Elev: 4,907 ft.
DeLorme Oregon Atlas & Gazetteer, p. 82.

6 John Day River at Muleshoe, Wheeler and Grant Counties

Highways 19 and 402 (the John Day Highway) from Service Creek in Wheeler County to Monument in Grant County follow the north side of the John Day and North Fork John Day Rivers for a 22 mile stretch. There are a number of public access points along this stretch, including the following: Donnelly, Service Creek, Muleshoe (a BLM campground), Lone Pine, Big Bend, and Monument Boat Ramp. You can spend a day dragonflying at these access points. To reach this area, turn north off Highway 26 just west of Mitchell onto Highway 207 and follow it 25 miles to Highway 19. Muleshoe campground is about 1.5 miles east along Highway 19.

A highlight dragonfly here is the Columbia Clubtail, which is known to occur in Oregon only in the John Day and Owyhee River basins. Other species to be expected include: Common Green and Blue-eyed Darners; White-belted Ringtail; Sinuous and Pale Snaketails; Western River Cruiser; Eight-spotted Skimmer; Western Pondhawk; Common Whitetail; Emma's Dancer; Northern, River, and Tule Bluets; and Western Forktail.

Ecoregion: Blue Mountains
Lat: 44.80734°; Long: 119.96697°; Elev: 1700 ft.
DeLorme Oregon Atlas & Gazetteer, p. 81.

7 Owyhee River at Rome, Malheur County

This is a good spot to find the Columbia Clubtail. The Owyhee and John Day River basins are the only known Oregon places where Columbia Clubtail can be found. Travel west from Jordan Valley on Highway 95 about 32 miles to Rome. A stop at the Owyhee River boat ramp on the east bank of the River can be productive. If you look north of Highway 95 at the bridge, you can see the beautiful "Pillars of Rome," a series of heavily eroded, white cliffs. To reach the gravel bar, where Columbia Clubtail may be found, continue west on Highway 95 across the River. Just past the Rome store, turn south on a dirt road and follow it about two miles and then west down to the west shore of the Owyhee River. Here the extensive gravel bars are a good place to find: Common Green and Blue-eyed Darners, Columbia Clubtail, Sinuous Snaketail, White-belted Ringtail, Western River Cruiser, Eight-spotted Skimmer, Variegated Meadowhawk, Emma's and California Dancers, and Tule Bluet.

Ecoregion: Basin and Range
Lat: N42.8060°; Long: W117.6133°; Elev: 3,395 ft.
DeLorme Oregon Atlas & Gazetteer, p. 75.

8 Mickey Hot Springs, Harney County

The Mickey Hot Springs site is located in the Alvord Desert, one of a number of interesting hot springs in this area. It is here that a number of southwest desert Odonate species find suitable habitat in Oregon at the northern limit of their ranges. To reach Mickey Hot Springs, take Folly Farm Road southwest from Highway 78 (65 miles southeast of Burns). It is 31.4 miles from Highway 78 to an east turn onto a dirt road, and another 6 miles to the well-marked Hot Springs. Exercise caution at hot springs in the area, as water can be very hot and capable of causing severe burns. In addition, signs warn of possible collapse of surface layers above underground hot springs. Western Rattlesnakes and fascinating lizards may also be seen in the Alvord.

At these hot springs, one can find some of the unusual species occurring in this part of Oregon including: Bleached and Hoary Skimmers, Desert Whitetail, Paiute Dancer, and Black-fronted Forktail. Other common species include Western Pondhawk, Eight-spotted and Twelve-spotted Skimmers, Variegated and Striped Meadowhawks, Western Red Damsel, and Western Forktail.

Ecoregion: Basin and Range
Lat: 42.6781°; Long: 118.3483°; Elev: 3,978 ft.
DeLorme Oregon Atlas & Gazetteer, p. 74.

Borax Lake, Harney County

9

The Alvord Desert in southeastern Oregon is a fascinating place and home to some interesting Odonates. This is the spot to see a number of southwest desert species that do not range into Oregon in any other place. Borax Lake is managed jointly by BLM and The Nature Conservancy and is reached via a dirt road running east along a power line near the intersection of Catlow Valley Road and Fields-Denio Road. Take this road and turn east at about 2.1 miles on a dirt road heading north and continue 2.6 miles to the lake. Note the warning sign on the wire gate into the lake. On a calm day, you may see water roil up from hot springs in the center of the lake.

Several unique Odonates can be found here including: Comanche, Bleached, and Hoary Skimmers; Desert Whitetail; Paiute Dancer; Alkali Bluet; and Black-fronted Forktail. In addition, Common Green, Blue-eyed, California, and Variable Darners; Dot-tailed Whiteface; Eight-Spotted and Flame Skimmers; Western Pondhawk; Striped and Variable Meadowhawks; Emerald and Spotted Spreadwings; Western Red Damsel; Tule Bluet; and Western Forktail may be encountered. These species may also be found at other hot springs in the area.

You may wish to take a side trip south on the Fields-Denio Road seven miles south of Fields where Cottonwood Creek crosses the road. This is the only place in Oregon where the Red Rock Skimmer has been found on a couple of occasions. Look for it perched on rocks in or along Cottonwood Creek.

Ecoregion: Basin and Range
Lat: 43.4403°; Long: 121.9321°; Elev: 4,063 ft.
DeLorme Oregon Atlas & Gazetteer, p. 74.

Twentymile Creek, Lake County

10

Twentymile Creek is a welcome sight in this hot and dry part of Oregon. This medium-sized creek is accessed 10 miles south of Adel along Twentymile Road. On the east side of the creek is a small spring where Aztec Dancer is found. This is the only spot to date where this species has been found in Oregon, and can be difficult to locate. This creek is very good dancer (*Argia*) habitat. In addition to Aztec Dancer, California, Emma's, Sooty, and Vivid Dancers are found here. Other species that may be observed include Common Green, Paddle-tailed, and Walker's Darners, Pacific Spiketail, Pale Snaketail, Flame Skimmer, Western Pondhawk, Striped Meadowhawk, California Spreadwing, Northern Bluet, and Western Forktail.

Ecoregion: Basin and Range
Lat: 42.0597°; Long. 119.9608°; Elev: 4,600 ft.
DeLorme Oregon Atlas & Gazetteer, p. 73.

11 Lake of the Woods, Klamath County

Lake of the Woods is located just south of Highway 140 approximately 42 miles east of Medford. The road along the east side of the lake provides access to a nice stop on the lake.

This is a popular and busy lake during the summer season and is the only known Oregon location for the Lance-tipped Darner. During the flight season, it can be common here—feeding in impressive swarms with a number of other darner species over the wet, grassy meadows along the lake shore. We have observed darners here catching and feeding on California Tortoiseshell butterflies when the populations of this butterfly are high. Other darners that may be seen here are Blue-eyed, Canada, Paddle-tailed, Shadow, and Variable. Additional species include: American and Mountain Emeralds, Chalk-fronted Corporal, Four-spotted Skimmer; Autumn, Striped, Band-winged, and White-faced Meadowhawks; Emerald and Northern Spreadwings, and Western Forktail (Gordon & Kerst, 2005).

Ecoregion: Eastern Cascade Mountains
Lat: 42.3767°, Long: 122.2026°, Elev: 5,016 ft.
DeLorme Oregon Atlas & Gazetteer, p. 21.

12 Jenny Creek, Jackson County

Jenny Creek is a wonderful small stream with rock and boulder substrate that can be accessed near the Pinehurst Inn along Highway 66 approximately 22 miles east of Ashland. The Creek has swift current, but there is also a slow section so both pond and stream species may be found here. This is excellent habitat for Walker's Darner and a location in Oregon where it is easy to find. Exercise caution as Western Rattlesnakes have been seen here. Other species that may be found here include: Common Green, Paddle-tailed, and Shadow Darner; Grappletail; Pacific Spiketail; Flame Skimmer; Autumn and Striped Meadowhawks; American Rubyspot; California Spreadwing; Vivid Dancer; Boreal and Familiar Bluets; and Pacific and Western Forktails.

Ecoregion: Eastern Cascade Mountains
Lat: 42.1177°; Long: 122.3672°; Elev: 3,403 ft.
DeLorme Oregon Atlas & Gazetteer, p. 21.

13 Crescent Creek, Klamath County

Crescent Creek is a beautiful mountain stream that can be accessed along Highway 58 approximately 4 miles west of the Crescent Cutoff. Park at the pullout on the southwest side of the bridge over the creek and walk up the dirt road. You will immediately begin to see Great Basin Snaketails and Emma's Dancers perched on the road and low vegetation. During the peak of the season, there are an amazing number of Great Basin Snaketail individuals and many copulating pairs here. Make your way down the bank to the creek and look for the snaketails perched on shrubs or the ground along the creek.

A number of other species may be observed here including: Paddle-tailed and Variable Darners; Pacific Spiketail; Ocellated Emerald; Spiny Baskettail; Four-spotted Skimmer; Black and White-faced Meadowhawks; Emerald, Northern, and Spotted Spreadwings; Boreal Bluet, and Swift Forktail.

Ecoregion: Eastern Cascade Mountains
Lat: 43.4823°; Long: 121.9011°; Elev: 4,620 ft.
DeLorme Oregon Atlas & Gazetteer, p. 44.

CASCADES LAKES HIGHWAY, KLAMATH AND DESCHUTES COUNTIES

14

The 78 mile Cascade Lakes Highway scenic drive extends from Highway 58 at the Crescent Creek Cutoff Road to Bend. Along the way, it passes by or near a dozen lakes and reservoirs and the wildflower-bordered Deschutes River, complete with its flyfishers. From the highway or from the lake shores, you can get breathtaking views of Mt. Bachelor, Broken Top, and the South Sister volcanic peaks. There are at least 25 Deschutes National Forest campgrounds along the way and Bend offers a full range of amenities. From south to north, 4 lakes offer good dragonflying: Davis, Crane Prairie, Hosmer, and Sparks (See also separate sections on nearby Gold Lake, Crescent Creek, and Todd Lake.)

To reach Davis Lake (Klamath County) from Highway 58, travel east 4 miles on the Crescent Cutoff Road and turn north on the Cascade Lakes Highway. In about 3 miles, take the turnoff to the West and East Davis Campgrounds. At the edges of Davis Lake's marshes, look for: Paddle-tailed and Variable Darners; Four- and Eight-spotted Skimmers; Striped, Saffron-winged, White-faced, and Black Meadowhawks; Northern, Emerald, and Spotted Spreadwings; Boreal Bluet; and Pacific Forktail.

Return to the Cascade Lakes Highway and travel north (crossing into Deschutes County) another 13 miles and turn east onto South Century Drive toward Sunriver. After about 4 miles, turn north on a dirt road toward the Brown Mountain Boat Landing. Stop when you see the dam at the south end of Crane Prairie Reservoir. In August, the population density of Great Basin Snaketails is amazing here! Look also for: Paddle-tailed and Variable Darners, American Emerald, Beaverpond Baskettail, Four- and Eight-spotted Skimmers, Chalk-fronted Corporal; Striped and White-faced Meadowhawks, Northern and Spotted Spreadwings, Emma's Dancer, and Pacific and Western Forktails.

Return to the Cascade Lakes Highway and travel north about 17 miles to the Hosmer Lake Boat Ramp, which is about 1.5 miles off the highway. This is a good place to find: Paddle-tailed and Variable Darners; American Emerald; Four-spotted Skimmer; Band-winged and Black Meadowhawks; Northern, Emerald, and Spotted Spreadwings; Northern and Boreal Bluets; and Western Forktail.

Return to the Cascade Lakes Highway and travel north and east about 9 miles, passing Elk and Devil's Lakes along the way. Pull off on the south side of the highway near Goose Creek and explore the marsh and bog at the northwest corner of Sparks Lake. Another mile east takes you to the campground. You can wade the marsh and meandering creeks (cold and fairly deep) toward the south end of Sparks Lake. The lake, bogs, and marshes support the following: Paddle-tailed, Variable, and Zigzag Darners; American, Ringed, and Mountain Emeralds; Hudsonian and Dot-tailed Whitefaces; Four, Eight, and Twelve-spotted Skimmers; Chalk-fronted Corporal; Striped and Black Meadowhawks; Northern and Emerald Spreadwings; and Taiga and Northern Bluets.

Ecoregion: Eastern Cascade Mountains
Davis Lake
Lat: 43.5894°; Long: 121.8520°; Elev: 4400 ft.
Crane Prairie Reservoirs
Lat: 43.7550°; Long: 121.7860°; Elev: 4450 ft.
Hosmer Lake
Lat: 43.9575°; Long: 121.7876°; Elev: 4990 ft.
Sparks Lake
Lat: 44.0301°; Long: 121.7385°; Elev: 5440 ft.
DeLorme Oregon Atlas & Gazetteer, pp. 44, 49, and 50.

15 Todd Lake, Deschutes County

Todd Lake is a very popular recreation spot only 25 miles from Bend off the Cascades Lake Highway. A 1.5 mile trail, accessed at the end of the spur road, circles the lake. A large wet meadow at 6,100 feet elevation on the sloping west bank of the lake is home to a good Black Petaltail population. This is a magical spot on a warm summer day and also features an impressive display of wildflowers in season. During the peak of the season you will not need to look hard for these interesting dragonflies as they may well land on you. They are attracted to light colors and commonly perch on the bleached logs lying around the meadow and are quite approachable. The attractiveness of these reflective surfaces is likely related to the need to remain warm as it can be cool at these elevations. Other species that may be observed along the lake shore or feeding in the open meadow include: Paddle-tailed and Variable Darners and American, Ocellated, and Ringed Emeralds.

Ecoregion: Eastern Cascade Mountains
Lat: 44.0251°; Long: 121.6839°; Elev: 6,100 ft.
DeLorme Oregon Atlas & Gazetteer, p. 50.

16 Camas Prairie, Wasco County

Camas Prairie is a wet prairie in northeastern Wasco County with small pools, ponded areas, and a stream running through the center of it. From Highway 26, take Highway 216 for approximately 2 miles east to Keeps Mill Road. Take Keeps Mill Road for 0.8 mile and turn left on a forest service road which dead ends at a trailhead. At the parking area, the prairie can be seen through the trees to the south.

This is the best location in Oregon to find the Subarctic Darner. The only other location in Oregon where this darner has been found is Little Crater Lake, but it has not been seen there for several years. Subarctic Darners are common at Camas Prairie in mid to late August. Other species that can be found at the prairie include: California, Canada, Paddle-tailed, and Variable Darners; Brush-tipped Emerald (uncommon), Four-spotted Skimmer, Common Whitetail; Black and Striped Meadowhawks; Emerald, Northern, and Spotted Spreadwings; Western Red Damsel; Boreal Bluet, and Western Forktail. The Hudsonian Whiteface is also common here and often lacks the characteristic red markings on the abdomen making them difficult to separate from the Belted Whiteface. Both species have unmarked abdomens along with single cells in the radial planate. Thus, it is necessary to examine the hamules to be certain of the species identification. Camas Prairie is a very pleasing spot to spend a warm summer afternoon.

Ecoregion: Eastern Cascade Mountains
Lat: 45.1385°, Long: 121.5611°, Elev.: 3,109 ft.
DeLorme Oregon Atlas & Gazetteer, p. 62.

GOLD LAKE/BOG/SALT CREEK – LANE COUNTY

17

Gold Lake lies north of Highway 58 just west of Willamette Pass on road, NFD 500. Salt Creek drains from Gold Lake and supports an amazing population of Grappletails. This is the best spot in Oregon to observe this interesting species as they perch on rocks, logs, and the bank above the bridge over the creek. Numerous other stream and lake species may be observed including: several species of darners and emeralds, Vivid Dancer, and Swift Forktail. The only record for Great Basin Snaketail west of the Cascade Mountain crest was from this spot.

Gold Lake Bog is a subalpine bog lying at 4,850 feet in elevation on the northeast side of Gold Lake. It is 463 acres in size with about 187 acres of marsh and bog. It can be reached by an easy hike on a trail (3677) at the end of Gold Lake Road (Forest Road 223) after passing through the campground. The hike takes about 15 minutes, and at an opening along the trail, the bog will be visible on the right down a slope. The bog itself consists of wet areas, ponds, streams, and surrounding forested slopes and is wet, awkward hiking. It supports an interesting variety of plants including wildflowers and six species of carnivorous plants, as well as interesting birds and insects. Gold Lake Bog is a marvelous place on a warm summer day with a carpet of sundew and great views of Diamond Peak. We have recorded 40 species of Odonata at these sites, and it is possible to see more than 20 species on a single trip. It has also produced several new Lane County records for Odonata, including the uncommon Zig-zag Darner and Brush-tipped Emerald. Some of the species you may see at the bog include: several species of emeralds; Dot-tailed, Crimson-ringed, and Hudsonian Whitefaces; Black Meadowhawk; Western Red Damsel; Taiga Bluet; and Sedge Sprite. We have also found nine species of darners here. The bog is a listed Federal Research Natural Area and should be treated gently.

Ecoregion: Western Cascade Mountains
Lat: 43.6386°; Long: 122.0409°; Elev: 4,850 ft.
DeLorme Oregon Atlas & Gazetteer, p. 43.

PORTLAND METROPOLITAN REGION, MULTNOMAH AND WASHINGTON COUNTIES

18

Dragonfly enthusiasts in the Portland area have a wonderful guide available to natural areas, "Wild in the City"—a guide to Portland's natural areas edited by Michael C. Houck and M. J. Cody (2000). In this guide, Ralph Thomas Rodgers recommends a number of spots in the metropolitan region as good spots for observing dragonflies and damselflies. Some of the recommended sites are shown on the site map on page 41:

Ecoregion: Western Interior Valleys (Willamette Valley)
Beggar's-Tick Marsh Wildlife Refuge, *DeLorme*, p. 60.
Lat: 45.4794°; Long: 122.5490°; Elev: 212 ft.
Oaks Bottom Wildlife Refuge, *DeLorme*, p. 60.
Lat: 45.4696°; Long: 122.6590°; Elev: 35 ft.

Jackson Bottom Wildlife Preserve, *DeLorme*, pp. 60 and 66
Lat: 45.5017°; Long: 122.9909°; Elev: 152 ft.
Fernhill Wetlands, *DeLorme*, p. 65.
Lat: 45.5092°; Long: 123.0913°; Elev: 158 ft.
Hedges Creek Marsh (near downtown Tualatin, behind Hedges Green Shopping Center)
Lat: 45.3818°; Long: 122.7683°; Elev: 125 ft.
Oregon Episcopal School Marsh, *DeLorme*, p. 60.
Lat: 45.4770°; Long: 122.7582°; Elev: 220 ft.
Koll Creekside Marsh, *DeLorme*, p. 60.
Lat: 45.4531°; Long: 122.7946°; Elev: 165 ft.

Freeway Lakes, Albany, Linn County

19

Freeway Lakes are a series of three borrow pits along I-5 3 miles south of Albany. Take the Highway 20 exit and travel south on Spicer Drive (east of I-5) which becomes Three Lakes Road. There is parking in the lot on the east side of the road. The lakes lie on either side of Three Lakes Road and I-5 and are connected by channels. This is a popular fishing spot and a good location to see the common valley Odonates. Among the species that we have observed are Common Green and Shadow Darners, Pacific Clubtail, Blue Dasher, Eight-spotted and Flame Skimmers, Cardinal Meadowhawk, Vivid Dancer, Tule Bluet, and Pacific and Western Forktails. You may also see an unusual specimen such as an Olive Clubtail, which has been photographed at these lakes. While watching quietly for Odonates, you could also catch sight, as we did, of a Green Heron catching a female darner which was ovipositing on a log.

Ecoregion: Western Interior Valleys (Willamette Valley)
Lat: 44.5917°; Long: 123.0583°; Elev: 301 ft.
DeLorme Oregon Atlas & Gazetteer, p. 53.

Sandpiper Pond, Eugene, Lane County

20

From downtown Eugene, follow W. 11th Ave. (Highway 126) west and turn north onto Bertelsen Road; after crossing Amazon Creek, turn east on Stewart Road; about ¼ mile east, park near a gate and walk north approximately ¼ mile looking for Sandpiper Pond to the west. This is a large pond created by the Oregon Department of Transportation for wetland impact mitigation and is part of the west Eugene wetland system.

We have recorded 39 Odonate species at this site, and it is considered one of the best representative sites in the Willamette Valley. It is the first Oregon site where a Wandering Glider has been observed mating. Other unusual sightings here are Cherry-faced and White-faced Meadowhawks, Spot-winged Glider, and Boreal Bluet.

Sandpiper Pond is a good low elevation site for darners, skimmers, meadowhawks, and spreadwings. Seven of the ten species of meadowhawks found in Oregon have been observed here. It is a good place to find: Common Green Darner; Flame, Widow, and Twelve-spotted Skimmers; five spreadwing species; and Tule Bluet. Both Beaverpond and Spiny Baskettails have also been observed here.

Ecoregion: Interior Valleys
Lat: 44.0530°; Long: 123.1584°; Elev: 395 ft.
DeLorme Oregon Atlas & Gazetteer, p. 47.

Mt. Pisgah Arboretum, Lane County

21

Mt. Pisgah Arboretum is located at the Howard Buford Recreation Area near Eugene. The Arboretum can be found by exiting I-5 at Exit 189 just south of Eugene and following directional signs to the park. The Coast Fork of the Willamette River flows adjacent to the park, and there is an old channel of the river that provides pond habitat. Thus, a diverse group of dragonflies and damselflies can be found here including both stream and pond species.

After parking in the lot provided (parking fee required), continue past park buildings on the Tom McCall Riverbank Trail along the river. The vegetation overhanging the river is an excellent spot to find the beautiful River Jewelwing and American Rubyspot and to observe their interesting mating and ovipositing behaviors. Perched in sunny spots on the ground near the river, one can find Pacific Clubtail and Emma's Dancer. Sinuous Snaketails are usually found perched on the ground in meadows near the river.

The Water Garden Trail leads to the Adkison Bridge which crosses an old river channel. This is a good spot to observe some pond species as well as jewelwings. Species commonly seen at this spot include: Common Green and California Darners, Blue Dasher, Common Whitetail, Cardinal Meadowhawk, and Tule Bluet. Darners and baskettails are also seen feeding in meadows at the park.

Ecoregion: Western Interior Valleys (Willamette Valley)
Lat: 44.0071°; Long: 122.9810°; Elev: 500 ft.
DeLorme Oregon Atlas & Gazetteer, pp. 42 and 48.

South Umpqua River, Myrtle Creek, Douglas County

22

The South Umpqua River is a convenient stop along I-5 just north of Myrtle Creek. Take Exit 108 off I-5 and cross the bridge over the South Umpqua River into Myrtle Creek. Shortly after entering town, turn north (left) onto Dole Road and follow it north along the east side of the river for about 1.1 miles and park in the gravel area to the west of the road. Here you will find river habitat, rocky shore, and adjacent short riparian vegetation to support a variety of dragonflies and damselflies during the peak season from June through August. Check the river bank, gravel bar, and nearby willows and the roadside from this point southward under the bridge.

You may see Common Green and Blue-eyed Darners; Sinuous and Bison Snaketails; Western River Cruiser; Eight- and Twelve-spotted, Widow, and Flame Skimmers; Blue Dasher; Cardinal Meadowhawk, Black Saddlebags; American Rubyspot; California, Emma's, and Vivid Dancers; and Pacific and Western Forktails.

Ecoregion: Western Interior Valleys (Umpqua Valley)
Lat: 43.0365°; Long: 123.3287°; Elev: 580 ft.
DeLorme Oregon Atlas & Gazetteer, p. 35.

23 Jackson County Sports Park, Medford, Jackson County

This park and nearby Hoover Ponds are good locations for many of the common pond species in the Rogue Valley. While the species seen here are not rare or unusual, the sheer numbers of individuals can be impressive. From Medford, drive north on Highway 62 (the Crater Lake Highway) approximately 5.5 miles and turn east on Highway 140 (the Lake of the Woods Highway). From Highway 62, travel just a bit over 2 miles and turn south into Jackson County Sports Park. You will see Richard's Pond on your left. Just across a paved road on the north of Richard's Pond is a larger pond with a trail around it. To the immediate west is Hoover Park with a series of three borrow pits fed by a small stream.

Many of the widespread species can be found at these ponds: Common Green, California, and Blue-eyed Darners; Beaverpond Baskettail; Flame, Eight-spotted, Twelve-spotted, and Widow Skimmers; Common Whitetail; Western Pondhawk; Blue Dasher; Variegated, Band-winged, and Cardinal Meadowhawks; Black Saddlebags; Emerald Spreadwing; Vivid Dancer; Tule, Boreal, and Northern Bluets; and Pacific and Western Forktails. Among unusual species to look for are Familiar Bluet; Black Spreadwing; and Spot-winged Glider, the latter during its irregular migrations.

Ecoregion: Western Interior Valleys (Rogue Valley)
Richard's Pond: Lat: N42.4172°; Long. W122.8055°; Elev: 1,417 ft.
DeLorme Oregon Atlas & Gazetteer, p. 20.

Illinois River Forks State Park, Cave Junction, Josephine County

Illinois River Forks State Park is located along Highway 199 at the south edge of Cave Junction. This is the spot where the east and west forks of the Illinois River meet. There are also some ponded areas from beaver dams and some backwaters, so it supports quite an interesting list of dragonflies and damselflies. There are large gravel bars and riffles where the Bison Snaketail can easily be observed. This is the best spot in Oregon to observe this unique and beautiful species. A number of other interesting Odonates can also be seen here including: California and Blue-eyed Darners; Pacific Clubtail, Western River Cruiser, Beaverpond and Spiny Baskettails; Eight-spotted, Flame and Widow Skimmers; Common Whitetail; Cardinal, Red-veined and Striped Meadowhawks; River Jewelwing; California, Emma's and Sooty Dancers; Northern Bluet, and Western Forktail.

Ecoregion: Western Interior Valleys (Rogue Valley)
Lat: 42.1599°, Long: 123.6573°, Elev.: 1,307 ft.
DeLorme Oregon Atlas & Gazetteer, p. 18.

25 Illinois River at Eight Dollar Mountain, Josephine County

Eight Dollar Mountain Road crosses the Illinois River about 3 miles west of Highway 199 (Redwood Highway). This is a refreshing stop on a hot summer day where temperatures can be above 100° F. There is easy access to the River with ample parking and primitive camping on the west side. There is a large population of Sooty Dancers, usually found perched on rocks near or in the stream. Other species observed here include: Pacific Clubtail, Flame Skimmer, Variegated Meadowhawk, Boreal and Northern Bluets, and California and Emma's Dancers.

In addition, there is a small pond upriver on the east side formed by a small stream that is cut off from the river. Shadow Darners and California Spreadwings have been seen here.

Ecoregion: Klamath Mountains (Siskiyou Mountains)
Lat: 42.2459°; Long: 123.6908°; Elev: 1,080 ft.
DeLorme Oregon Atlas & Gazetteer, p.18.

Darlingtonia Fens, Josephine County

26

The Illinois Valley in southeast Oregon has a unique type of wetland. Fens in the Illinois Valley are formed where alkaline groundwater finds its way to the surface through fractures in serpentine rock. At these sites, the insectivorous California Pitcher Plant (*Darlingtonia californica*) thrives along with a unique plant community. The wildflowers and butterflies here can also be impressive in the spring. These sites all appear to have good populations of the Black Petaltail. The larvae live in burrows in the wetlands and are thought to have a five-year life cycle. It is a unique sight to see one of these primitive dragonflies perched on a "pitcher" of this plant. One may also encounter other Odonates here as there are often pools or small streams draining the area. We have seen California Darner, Flame Skimmer, Pacific Spiketail, and Vivid Dancer in the fens.

Lone Mountain Road out of O'Brien has a number of fens along it varying in size from very small to perhaps two acres. These can be found along the road from 3 miles to 6.3 miles from O'Brien. A large fen can be found just north of the road after crossing Whiskey Creek at mile 5.6. There are also two large fens that can be reached by taking Eight-dollar Mountain Road approximately 4 miles north of Kirby. A developed BLM site, complete with a boardwalk and interpretive information, is easily found by following the signs on Eight Dollar Mountain Road. To reach Days Gulch Fen, continue on Eight Dollar Mountain Road, cross the Illinois River, and proceed on NF-029. The fen is located 2 miles beyond the river on the north side of the road.

Whiskey Creek Fen: Lat: 42.0226°; Long: 123.7729°; Elev: 1,710 ft.
Eight Dollar Mountain Fen: Lat: 42.2320°; Long: 123.6585°; Elev: 1,600 ft.
Days Gulch Fen: Lat: 42.2297°; Long: 123.7027°; Elev: 1,550 ft.
DeLorme Oregon Atlas & Gazetteer, p. 18.

Arizona Beach State Park, Curry County

27

Arizona Beach State Park is located 12 miles south of Port Orford on Oregon's ruggedly beautiful south coast. Just east of Highway 101 in the park, there are a small pond and wetland that host a diverse Odonate population for a coastal location. At this site, California Spreadwing was found in Curry County for the first time. The darners are represented by Common Green, Blue-eyed, Paddle-tailed and Shadow. Other species that may be seen include: Eight-spotted, Twelve-spotted, Flame, and Widow Skimmers; Blue Dasher; Cardinal and Variegated Meadowhawks; Black Saddlebags; Boreal, Northern, and Tule Bluets; Pacific Forktail, and Northern Spreadwing.

Ecoregion: Coast Range
Lat: 42.6167°; Long: 124.3947°; Elev: 22 ft.
DeLorme Oregon Atlas & Gazetteer, p. 25.

28 Garrison Lake, Port Orford, Curry County

Garrison Lake is located on the western edge of Port Orford on the south Oregon Coast. There are two easy spots to access the lake; a boat ramp at the west end of Twelfth Street, and at Paradise Point on County Road 244. The lake shore around the boat ramp presents the best habitat. The cool coastal environment limits the diversity of Odonata that are found. Common species observed include: California, Blue-eyed, Paddle-tailed, and Common Green Darners; Eight-spotted Skimmer; Cardinal, and Variegated Meadowhawks; Northern and Tule Bluets; and Pacific Forktail.

Ecoregion: Coast Range

Paradise Point: Lat: 42.7600°; Long: 124.5118°; Elev: 22 ft.

Boat Ramp: Lat: 42.7478°; Long: 124.5086°; Elev: 33 ft.

DeLorme Oregon Atlas & Gazetteer, p. 24.

29 Ten Mile Lake, Coos County

This is a large lake and can be visited at several spots including Ten Mile County Park in Lakeside south of Reedsport. The park is just over a mile from Highway 101 off 11th Street. Common species that can be found in the cooler coast climate can be seen here. These include Common Green, California, and Blue-eyed Darners; Blue Dasher; Cardinal Meadowhawk; Eight-spotted Skimmer; Pacific Forktail; and Tule Bluet.

Ecoregion: Coast Range

Lat: 43.5735°; Long: 124.1713°; Elev: 18 ft.

DeLorme Oregon Atlas & Gazetteer, p. 32

30 Sunset Beach State Recreation Area, Clatsop County

(Submitted by Mike Patterson)

To reach this north coast site, travel about 4 miles south on Highway 101 from Warrenton (or 4 miles north from Seaside) and turn west onto Sunset Beach Lane. Follow it to the parking lot. This 120-acre park is the western terminus of the Fort-to-the-Sea trail from the Lewis and Clark Historical Park. You won't be disappointed at Sunset Beach. From atop the sand dunes you can see the Pacific Ocean and look north and get an excellent view of Cape Disappointment, Washington, on the far side of the Columbia River mouth. Take the loop trail through the sand dunes to the interdunal lakes which are bordered by shorepine, willows, and rush/sedge wetlands. Eighteen species of Odonates have been recorded at the park or nearby.

Among species found are: Common Green, Paddle-tailed, California, and Blue-eyed Darners; American Emerald; Dot-tailed Whiteface; Variegated, Cardinal, Red-veined, Striped, and Autumn Meadowhawks; Northern Spreadwing; Tule and Northern Bluets; and Pacific Forktail. Spot-winged Glider has also been recorded here.

Ecoregion: Coast Range

Lat: 46.0993°; Long: 123.9360°; Elev: 29 ft.

DeLorme Oregon Atlas & Gazetteer, p. 70.

Map Site 3: Enterprise Wetland

Map Site 4: Anthony Lake

Map Site 5: Magone Lake

Map Site 8: Mickey Hot Spring

Map Site 9: Borax Lake

Map Site 15: Todd Lake

Map Site 16: Camas Prairie

Map Site 19: Freeway Lakes

Map Site 24: Illinois River

Map Site 26: *Darlingtonia* Fens

Examples of prime dragonflying habitat across Oregon

III. Species Accounts

The species accounts are presented with dragonflies first, followed by damselflies. The species are generally in family and species checklist order, but the order was altered in a few cases to group similar species together, which allows for easier comparison of similar species. For example, Blue Dasher and Western Pondhawk are presented on sequential pages, and petaltail, cruiser, and spiketail families, each represented by a single black and yellow species, are lumped together.

The information presented in this field guide is up to date through the 2009 field season. As Odonate enthusiasts explore Oregon's diverse habitats, our knowledge of the fauna, distributions, and flight seasons will continue to evolve.

Species Descriptions

We first introduce the family and genus with a general description and overview. In some cases, we also include a figure or an identification chart, which illustrate differences among or between similar species. Throughout the species descriptions, the colored marginal bar denotes the family name.

Descriptions are provided for males and females of each species, using common terms rather than morphological terms as much as possible. We have not provided a glossary but any technical terms used in the guide can be found in the morphology section or figures. Both English and scientific names are provided but English names are used throughout the text. Males are the most commonly encountered sex by observers as they are more common at water. Thus, they are listed first in the species descriptions.

Photographs, Figures, and Identification Charts

Opposite the species description, we provide photographs that illustrate characters used in identification of species. In many cases, these are side views showing patterns on the thorax and abdomen, and eye, leg, and facial colors. Top views are provided

Eight-spotted Skimmer male

where identification can be based on wing patterns. For species where a lateral view is important, photographs have been rotated to better facilitate comparison between species. In each instance, we have selected a natural or posed photograph that provides you with the best view of key field marks. For groups such as darners, spiketails, and emeralds, note that these Odonates have long legs and the natural perching position is to hang vertically. For some species, we provide photographs of more than one color form for females, but we have not attempted to cover all color forms. We have used inset photographs and figures to highlight key field marks for selected species, families, and genera. For most groups, we have included an identification chart that illustrates and annotates key features.

When observing Odonates in the field, note as many characters as you can for comparison to photos, illustrations, and species descriptions. Note features such as overall color and colors or patterns on the wings. If the wing has color, look for distinct spots, a wash of color at the base, or colors of veins on the leading edge of the wing. Check for the size, color, and shape of stripes on the thorax, and spots or stripes on the top and sides of the abdomen. Look for the color of the face, eyes, legs, and terminal appendages. All of these characters can be diagnostic for a species identification.

Illustrations

The drawings of body parts are used to help identify species that have very similar markings. This usually entails having the insect in hand and using a hand lens to magnify the subject parts. We have not shown all the hairs, which can be abundant and sometimes obscure a clear view of the parts. We have not shown parts best identified with a microscope. Such detail is probably beyond the interest of most users of this field guide.

Range Maps, Distribution and Habitats

Range maps were developed from our personal experience, county records on Odonata Central, and the excellent work on Oregon Odonata published in the Bulletin of American Odonatology by Jim Johnson and Steve Valley (http://odonata.bogfoot.net). This issue of the bulletin is not referenced under each species description to avoid repetition, but it is a major reference upon which this section of this field guide is based. Johnson and Valley's work not only outlined the current knowledge of the distribution of the Odonata in Oregon but listed six additional species that should be expected to be found in Oregon. To date, three of those six species have now been found in Oregon.

The shaded areas of the range maps represent our current understanding of the distribution of each species in Oregon. The dark lines are ecoregion boundaries as shown in detail on the ecoregion map (see map on page 40). When identifying Odonates in the field, first consider species that are within the distribution range of species at a given site. However, extensions of known ranges are found every year. Thus, observation of a species outside its known range is always a possibility. The general North American distribution as well as the Oregon distribution is referenced for each species. General habitat preferences and elevation limits are also noted.

Flight Dates

The flight periods for species are taken from Oregon records maintained by Jim Johnson for the Oregon Dragonfly Survey. These dates are the earliest and latest dates recorded over many years. Thus, one should not expect to find these species flying every year on the early and late dates. In addition, there is considerable distance from north to south in Oregon, and sea level to high elevation mountain habitat. These variables have the expected impact on emergence periods: expect earlier emergence in the south and at lower level elevations.

Size

The size given for each species represents body length which was developed from specimens in personal collections, and from west coast references: Paulson (1999), Paulson (2009), Manolis (2003), and Cannings (2002). For each species a size range in millimeters is given as well as a silhouette showing an average length. There are size variations within a species depending on sex, time of season, and geographic distribution, in addition to inherent variability. The relative body length data and terms (long, medium, short) provide the reader with a relative size comparison among a set of similar species (family, genus, or even among similar families—like the spiketails and cruisers).

Thus, in the meadowhawk genus, the Autumn Meadowhawk is "short" (31–37 mm) when compared with the "long" Variegated Meadowhawk (40–42 mm). These relative length terms do not apply when comparing across groups. The "very short" California Darner (57–60 mm) is indeed a short darner, but is much longer than the relatively "long" Mountain Emerald (47–52 mm).

The range of body lengths and the silhouette figures are more useful tools for comparisons between the species within families or other subgroups. The size ranges listed are for both males and females. Females tend to have shorter, thicker abdomens and are shorter than males in total body length. It is important to recall that size can be a good character where there are large discrepancies in size, but size is variable.

The Wonder of Dragonflies

Black Petaltail male on dried California Pitcher Plant

A. Dragonflies

Sixty-three (63) species of dragonflies are presented in the following pages. The dragonflies are in the insect order Odonata and comprise the suborder Anisoptera. The damselfly species' presentations follow the dragonfly presentations. Damselflies comprise the other suborder, Zygoptera.

Dragonflies have eyes that touch or are separated by no more than one eye width. They have a robust thorax and abdomen. Their forewings and hindwings are shaped differently. The hindwing is expanded at the rear of the wing. At rest, dragonflies hold their wings horizontally perpendicular to the body. Dragonflies have two upper appendages (cerci) and one lower appendage (epiproct). See the morphology section and suborder key in the introduction for comparison of dragonflies and damselflies (see Figure 2, page 22).

The sixty-three dragonfly species are represented by seven families in Oregon: petaltails (Petaluridae), spiketails (Cordulegastridae), cruisers (Macromiidae), darners (Aeshnidae), emeralds (Corduliidae), clubtails (Gomphidae), and skimmers (Libellulidae).

Petaltails, Family Petaluridae

Black Petaltail larva

Black Petaltail burrow

The members of the Petalurid family are ancient insects with relatives that lived 150 million years ago during the Jurassic Period, when dinosaurs roamed the earth. The fossil record shows that the Petalurids were the dominant dragonflies on earth at that time. The larvae of this family are quite unusual—inhabiting seeps, bogs, and fens and living in burrows or under debris such as leaves.

The western species, *Tanypteryx hageni*, is found from British Columbia to Northern California. The common name relates to the large terminal appendages on some species of the family. There are eleven species of petaltails in the world with a single species of a different genus, *Tachopteryx thoreyi*, the Gray Petaltail, found in eastern North America, and a member of the same genus, *Tanypteryx pryeri*, found in Japan (Cannings, 2002, Manolis, 2003). The remaining eight species are all found in Australia, Argentina, Chile, and New Zealand.

♂ terminal appendages (bottom view)

Identification Chart D • Petaltail, Spiketail, and Cruiser

Species (size)	Eyes (top view)	Thorax (side view)	Abdomen (top view)
Black Petaltail ***Tanypteryx hageni*** Long 53–55 mm	 Separated, dark brown	 Yellow spots	 "U"-shaped yellow spots
Pacific Spiketail ***Cordulegaster dorsalis*** Very Long 72–78 mm	 Barely touching, aqua blue	 Two yellow stripes	 Yellow spots
Western River Cruiser ***Macromia magnifica*** Very Long 70–75 mm	Touch along seam	One yellow stripe	Light yellow spots

Western River Cruiser female

Families: Petaluridae, Cordulegastridae, and Macromiidae

Habitat and notes
Hillside seeps, *Darlingtonia* fens. Very long pterostigma.
Small streams, trickles, roadside ditches. Female—long ovipositor.
Medium to large rivers.

Note: Identification Chart.

Three species from three separate families are presented for comparison; Black Petaltail, Pacific Spiketail, and Western River Cruiser. Each family is represented in Oregon by a single species. While not closely related, all are large species, with black and yellow patterning. The eye color and position, markings on the side of the thorax, and markings atop the abdomen help separate these species in the field. Furthermore, they frequent different habitats, but habitat use should not be substituted for accurately noting field marks.

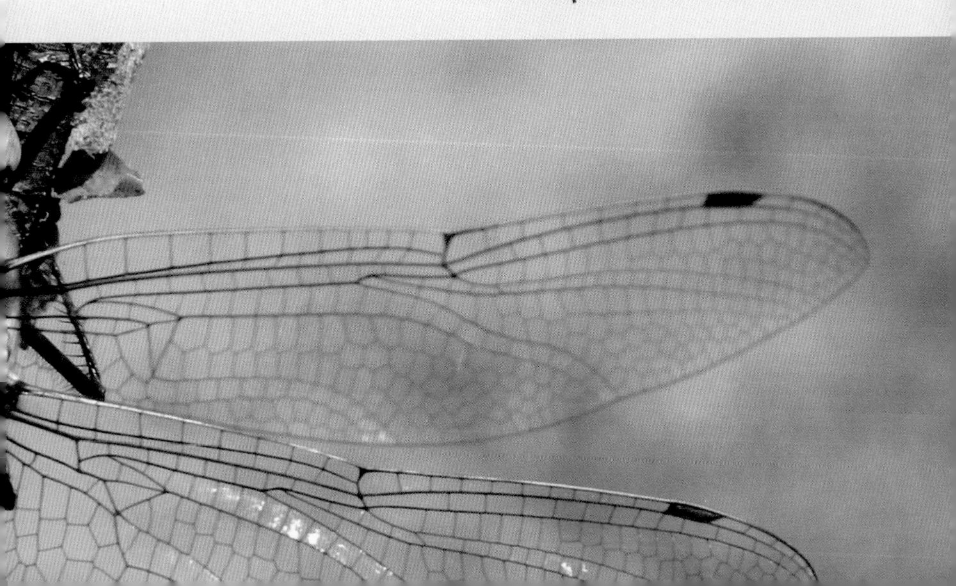

Black Petaltail

Tanypteryx hageni

This is a large black dragonfly with yellow spots on the sides of the thorax and yellow, thick, "U"-shaped spots atop the middle abdominal segments (S2–7). Its face is light yellow, and its dark brown or blackish eyes are separated. The clear wings have a uniquely long and narrow pterostigma. It has black legs. The male's appendages are broad and shaped like leaves or flower petals, thus its English name. Seen from above, the appendages are flat and angle outward (see illustration on page 59). The female and male are colored and patterned similarly, and the female exhibits a stout, blunt-tipped ovipositor which curves upward.

The Black Petaltail can be found in saturated, hillside seeps and *Darlingtonia* fens, where sheet water flows over the surface. The female inserts her eggs into mucky, mossy material. The larvae burrow 5–6 inches into the wet mud. The burrow generally runs down a few inches, and then curves upward a bit near the end. Larvae wait at the burrow entrance to ambush prey. These burrows are often concentrated in a very small area.

The Black Petaltail is found only from British Columbia to northern California. Once thought to be rare in Oregon, this species has now been located in over forty locations in the Cascade and Siskiyou Mountains, and most recently in the northern Coast Range.

Black Petaltails perch horizontally on logs, rocks, roads, and bare ground. They also hang vertically on tree trunks, snags, and even people. They are very unwary and can be observed closely.

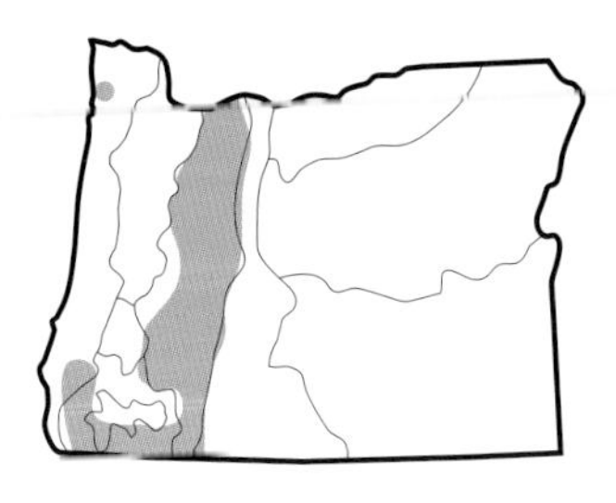

The adult flight season extends from late May to early September.

JAN	FEB	MAR	APR	MAY	JUN	JUL	AUG	SEP	OCT	NOV	DEC

Black Petaltail male

Black Petaltail female

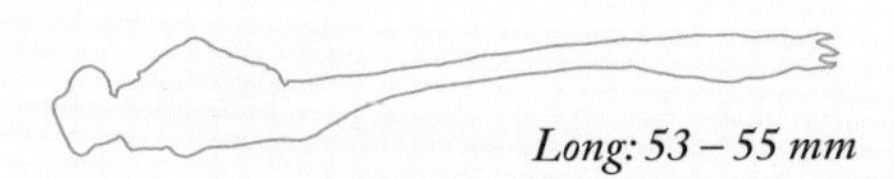

Long: 53 – 55 mm

Black Petaltail male

Western River Cruiser female

Spiketail Family, Cordulegastridae

The spiketail family gets its name from the female's large, pointed ovipositor (see photo to right). *Cordulegaster* is the single genus in this family in North America. Oregon has only one of the nine North American species of this family, but what a dragonfly the Pacific Spiketail (*Cordulegaster dorsalis*) is! Members of the family are large and have dramatic black and yellow colors (see Identification Chart D). It is especially impressive to see the males patrolling along very small streams. The adults have very long legs and, when perched, hang from vegetation nearly perpendicular to the ground. They can sometimes be found hanging in vegetation before the heat of the day.

The life cycle takes five years (Ken Tennessen, pers. comm.), which indicates that they must find refuges in some of the small streams which are nearly dry during part of the year.

Spiketail larva • photo by Giff Beaton

♀ Pacific Spiketail ovipositor

The Wonder of Dragonflies

Pacific Spiketail female (natural pose, perching)

Pacific Spiketail

Cordulegaster dorsalis

The Pacific Spiketail is a large, beautiful dragonfly. The thorax is dark brown with two bold yellow side stripes and two yellow frontal stripes. The abdomen is black with closely paired yellow spots on the top of segments 2–9. On close examination you can see a slight black line separating the yellow spots on some segments. Its small, blue eyes barely touch on top of the head, and its face is a pale, creamy color. It hangs up vertically or at a 45-degree angle when perched. The male and female are colored and patterned similarly, but the female's "spike" extends well beyond the abdominal tip. When females oviposit, they fly with the body held vertically and use their "spike" to punch eggs into wet mud or sand in an up and down manner, much like the needle on a sewing machine. They can be observed exhibiting this behavior as they move up and down the stream ovipositing in suitable spots. Males fly a foot or so above the water near the shoreline looking for females. If one waits, they will usually fly back the other direction some time later. It is an impressive sight to see this big, bold dragonfly patrolling a very small stream. Spiketails have been known to eat bees and wasps.

Pacific Spiketails inhabit a variety of streams from mountain trickles to larger creeks from sea level to 7,000 feet in elevation. This western species' range extends from southern British Columbia into southern California. In Oregon, it inhabits streams along the coast and mountainous regions, including foothills. It is scarce in the Western Interior lowlands below 500 feet and along the Columbia River in the north-central region. Several mountain streams in southeastern Oregon are inhabited by Pacific Spiketail. The larvae are shallow burrowers in streams—sitting with the eyes just out of the sediment, watching for prey.

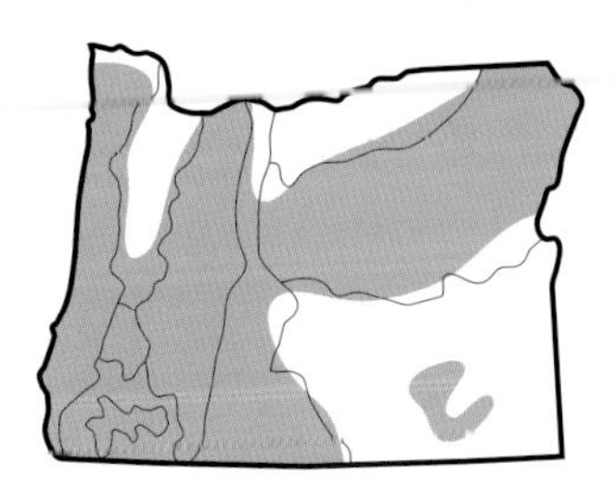

The adult flight season is from early June to late September.

JAN	FEB	MAR	APR	MAY	JUN	JUL	AUG	SEP	OCT	NOV	DEC

Pacific Spiketail male

Pacific Spiketail female

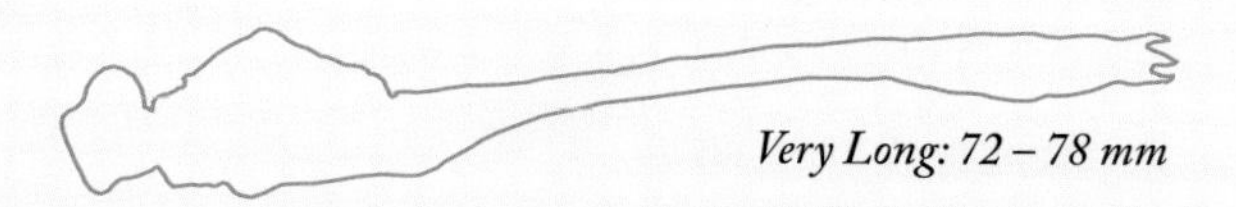

Very Long: 72 – 78 mm

Variegated Meadowhawk male

Flame Skimmer male

Cruiser Family, Macromiidae

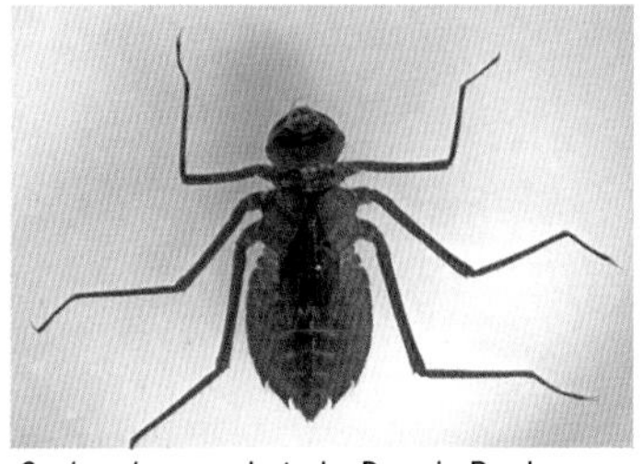

Cruiser larva • photo by Dennis Paulson

The cruisers are so named due to their rapid flight up and down streams, flying just above the water and often near the bank searching for females. They are also seen feeding in open areas, such as roads and nearby meadows. Worldwide there are 121 cruiser species. There are only two North American genera in the family, *Macromia* and *Didymops*. Some authors place this group as a subfamily of the emerald or skimmer families. Only one of nine North American species is found in Oregon, the Western River Cruiser (*Macromia magnifica*). This is another family of impressive, large, black and yellow dragonflies (see Identification Chart D). Our western species is appropriately named *magnifica* as it truly is a dazzling dragonfly. Cruisers have very long legs and hang vertically when perched, but they are seldom observed in a perching position. However, they may be found perched occasionally during cooler weather. The larvae are classified as "sprawlers": sitting in debris awaiting prey.

Western River Cruiser

Macromia magnifica

This large, dark and yellow dragonfly is a magnificently colored, strong-flying cruiser. Its eyes are pearly-gray and touch each other along a seam at the top of the head. The face is pale with a black cross-line. Its thorax is dark brown with one broad, yellow side stripe and short, yellow, frontal stripes. The abdomen is black with wide, yellow, squarish bands across the top of each segment, with the largest band on segment 8 and narrowing beyond. The end of the male's abdomen is slightly clubbed. This species has very long, black legs. The male and female are colored and patterned similarly, but the female's abdominal bands are larger and paler than the male's, and her abdomen is not clubbed.

The Western River Cruiser's North American range is western: from southern British Columbia southward into Mexico and east into Arizona. In Oregon, it is not known from the coast, and it is not common in the northern Willamette Valley. It can be found along permanent streams and is most common along larger rivers up to 4,000 feet in elevation. It is very common at the Three Forks of the Owyhee River in Malheur County.

This species is an extremely fast flier and males range long distances up and down rivers on patrol flying near the water surface. The female flies rapidly while laying eggs by tapping the water's surface. This species hangs vertically in trees when perched. It can be found foraging over meadows and fields, often feeding at tree-top level.

The flattened larvae are sprawlers and hunt by sitting on the stream bottom. They have a horn in front of the eyes and have very long legs (Behrstock, 2008).

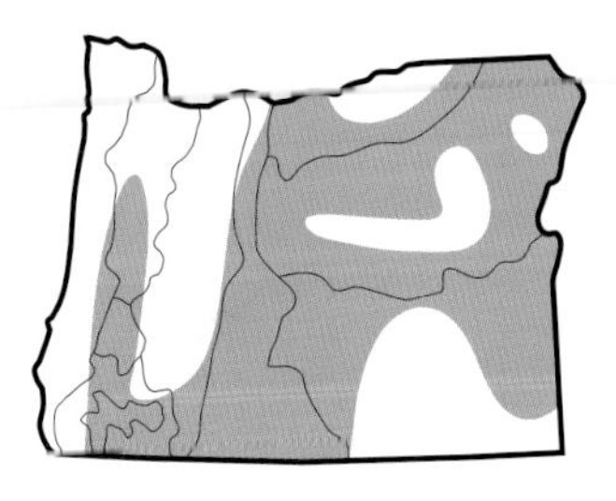

The adult flight season is fairly short, from early June to late August.

JAN FEB MAR APR MAY JUN JUL AUG SEP OCT NOV DEC

Western River Cruiser male

Western River Cruiser female

Darners, Family Aeshnidae

Members of this family are some of our largest dragonflies and are commonly called darners. The English name derives from the folk name, darning needle, which was used for dragonflies (Montgomery, 1972). These are large blue, brown, and green dragonflies. Female darners have both green and blue color forms and can be difficult to identify. They can often be seen in our backyards, gardens, and virtually any open space—where they fly high and dart up and down acrobatically catching insect prey. On hot days, they never seem to land but are in constant motion. They do not perch as some dragonflies do, but rather hang in vegetation by their long legs, especially during cooler weather.

Around lakes, streams, and ponds, the males are seen patrolling the shoreline in search of females. They often search the rushes and reeds along the banks, and it is common to hear the rustling of their wings as they work their way through emergent reeds. The rustling sound is greatly magnified when a female is located and copulation takes place. The coupled pair will usually fly up out of the vegetation often in the wheel position, fly around a bit, and land in nearby vegetation. Female darners have sharp ovipositors which are used to insert the eggs into vegetation or soil. Females can be seen clinging to emergent vegetation or partially submerged logs, a favorite oviposition site. The abdomen moves around over the submerged surface where the eggs are inserted. At times, they can also be seen ovipositing above the waterline along mountain lakes and streams. The eggs will be submerged when water levels rise. While females of most species oviposit without guarding by males, the Common Green Darner is often seen with the male contact guarding. Contact guarding is the term used when the male remains attached to the female during oviposition. In this manner, the male defends the female from fertilization by other males.

Darner larva

Larvae of the darners are voracious feeders that climb on submerged vegetation, logs, and other surfaces. Small larvae of many darners are banded with light and dark colors which disappear in later stages. This is postulated to protect larvae from predation by helping them blend into surrounding colors in the environment (Corbet, 1999). They seem to particularly like the rough surface of logs. Almost any invertebrates—including other Odonate larvae, tadpoles, and even small fish—can be prey for these efficient predators. Darner larvae are very streamlined in shape and seem especially good at propelling themselves by expelling bursts of water from the rectal chamber.

The Common Green Darner, *Anax junius*, is known to migrate, and large numbers of this species have been observed moving down the east coast. In some cases, millions of individuals have been observed in migrating swarms of dragonflies (Corbet, 1999). These swarms of dragonflies can be an important source of food for birds. Purple Martins, Swainson's Hawk, and Merlin are all documented as taking large numbers of dragonflies (Corbet, 1999).

Thirteen of North America's forty-one species of darners have been recorded in Oregon. Oregon's species belong to three different genera: the green darners, Genus *Anax,* (one species), are quite large and are named for the unmarked green thorax. They have a "bull's-eye" mark atop the head (see Identification Chart E). The mosaic darners, genus *Aeshna,* (ten species), typically have blue or green thoracic stripes and blue or green spots atop a dark abdomen. They are named after these markings, which are colored like brilliant mosaic tiles. They lack a tubercle under abdominal segment 1 and have a T-spot on top of the head (see Figure 3, page 74).

The Neotropical darners, genus *Rhionaeschna,* (two species), are also colored with mosaic blue and green patterns, but with a tubercle under abdominal segment 1 and a T-spot (see Figure 3).

Key features of the darners are illustrated on three identification charts: Green and Neotropical Darners (Chart E), Mosaic Darners with paddle-tailed upper appendages with a trailing spike (Chart F), and Mosaic Darners with simple upper appendages (Chart G).

Lance-tipped Darner male

Identification Chart E • Darners, Genera: Green Darners, *Anax*, and Neotropical Darners, *Rhionaeschna*

Green Darners, Genus *Anax* (solid green thorax; no side stripes):

Species (size)	Face; Head (top view)	Thorax (side view)	Appendages ♂ (side view)	Appendages ♀ (side view)
Common Green *Anax junius* Very Long 72–78 mm	No facial line; No "T-spot"; "bull's-eye" diagnostic	Solid green with no side stripes diagnostic	S10 (black)	

Neotropical Darners, Genus *Rhionaeschna*:

Species (size)	Face; Head (top view)	Thorax Stripes (side view)	Upper Appendages ♂ (side view)	Appendages ♀ (side view)
Californica *Rhionaeschna californica* Very Short 57–60 mm	Thin black line; T-spot very thick; stem widens at base	Narrow, pale blue	Simple, leaf-like; S10 (blue spot on top)	
Blue-eyed *R. multicolor* Short 62–69 mm	No facial line; bright blue eyes; T-spot with very thin crossbar	Broad and straight-sided; bright, sky blue	Forked diagnostic; S10 (blue spot on top)	

Figure 3: Darner Features

Green Darners, Genus *Anax*

Mosaic and Neotropical Darners, Genera *Aeshna* and *Rhionaeschna*

Neotropical Darners, Genus *Rhionaeschna*; tubercle (bump) under abdominal segment 1

(top of head)

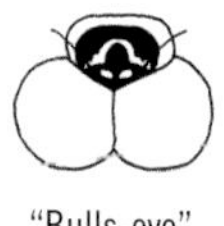

"Bulls-eye"

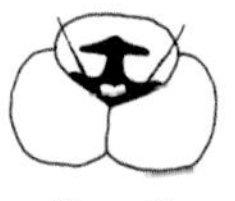

"T-spot"

(side view)

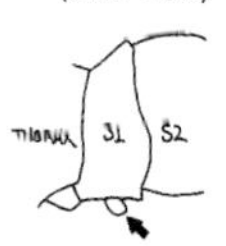

Identification Chart F • Mosaic Darners, Genus *Aeshna*, with Paddle-type Upper Appendages

Species (size)	Head: Facial Line; T-spot (top of head)	Thorax Stripes (side view)	Appendages ♂ (side view) Abdominal Notes	Appendages ♀ (side view)
Lance-tipped ***Aeshna constricta*** Long 65–74 mm	None  T-spot stem widens at base	Broad; front stripe with light indentation 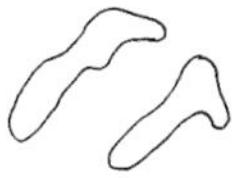 Both with rear extensions; pale blue	Upper appendages: very long, narrow; lance-like  S10 top, blue spot	
Paddle-tailed ***A. palmata*** Long 64–77 mm	Yes  T-spot: narrow stem; straight-edged crossbar	 Straight, sometimes wavy	Upper appendages: very broad paddle; spike extends beyond tip  S10 top, blue spot	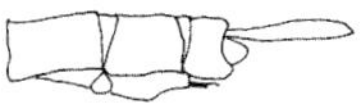
Shadow ***A. umbrosa*** Long 65–73 mm	Black or brown line  T-spot: short, thin stem; straight-edged crossbar	 Both with rear extensions	Upper appendages: long spike  S4-6 underside with paired spots; S10 top, no blue spot	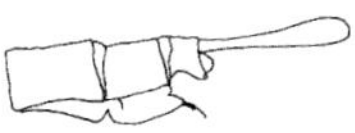
Walker's ***A. walkeri*** Medium 67–70 mm	Yes, slight brown line  T-spot: crossbar arms angled back	 Straight with black border; white or very pale	Upper appendages: extremely broad paddle, very short spike  S10 top, no blue spot	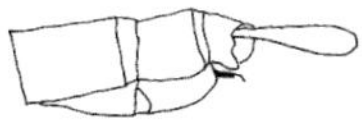

Identification Chart G • Mosaic Darners,

Species (size)	Head: Facial Line; T-spot (top of head)	Thorax Stripes (side view)	Appendages ♂ (side view) Abdominal Notes	Appendages ♀ (side view)
Canada *Aeshna canadensis* Long 66–73 mm	None Widens at base	Front deeply notched Rear "shoe"-shaped	Small teeth on top edge Paired spots underneath abdomen	
Variable *A. interupta* Medium 67–71 mm	Yes Stem and crossbar thick; widened base	 Dashes: very thin or broken	Simple, upturned lobe	
Sedge *A. juncea* Medium 66–71 mm	Yes; thick, black line  Crossbar with blurred front edge	Very broad, straight  Bordered in black	Narrow, pointed  Paired spots underneath abdomen	

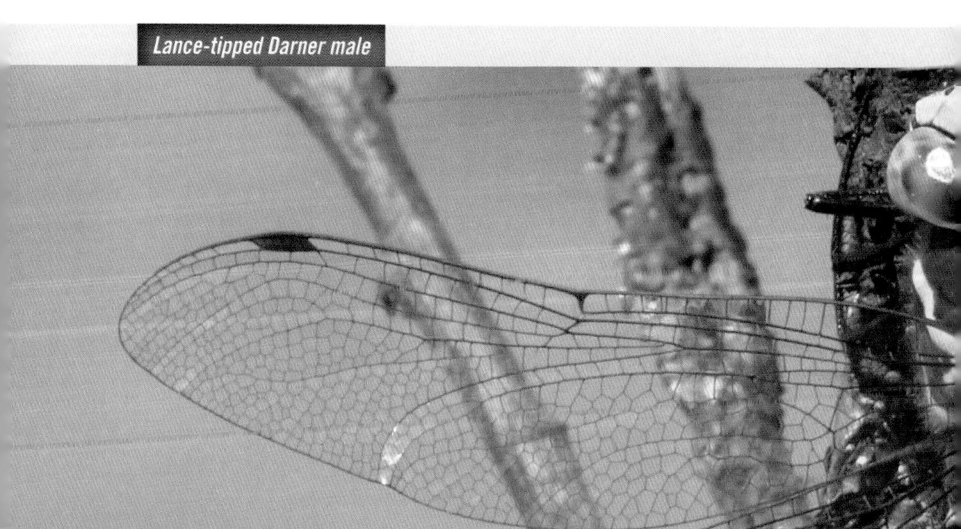

Lance-tipped Darner male

Genus *Aeshna*, with Simple Upper Appendages

Species (size)	Head: Facial Line; T-spot (top of head)	Thorax Stripes (side view)	Appendages ♂ (side view) Abdominal Notes	Appendages ♀ (side view)
Zigzag *A. sitchensis* Short 55–68 mm	Yes Base convex upward, diagnostic	Front thin, "zigzag" Rear "T-bone"-shaped, diagnostic	Petal-shaped	
Subarctic *A. subarctica* Short 63–69 mm	Yes Narrow, straight stem	Both bent in middle Front stripe rearward extension	Petal-shaped, very hairy, tip pointed downward Paired spots underneath abdomen	
Black-tipped *A tuberculifera* Very Long 71–78 mm	None Crossbar ends tipped downward	Broad front stripe; rearward extension	No spot atop S10, diagnostic Tubercle underside upper appendage	

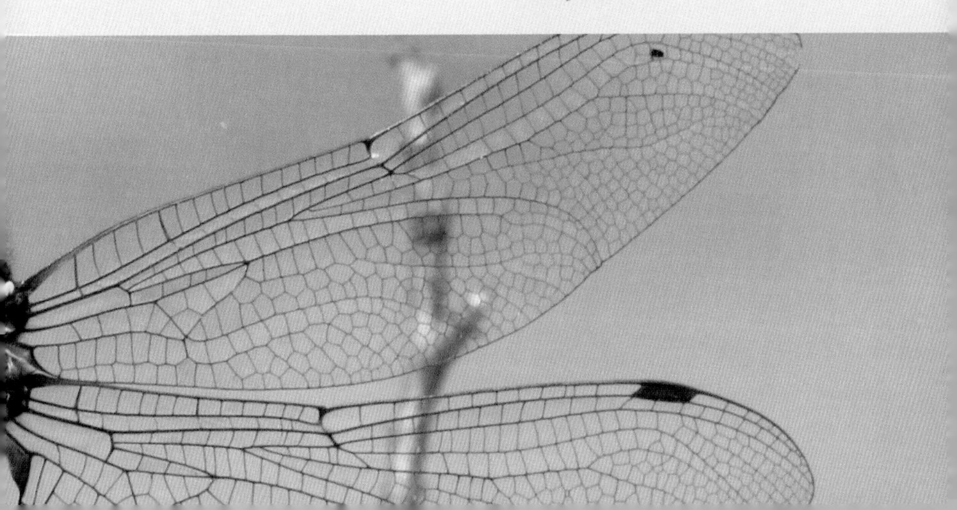

Darners

Common Green Darner

Anax junius

This is a large, truly magnificent dragonfly with bold green and blue markings. Males have a solid green thorax, green eyes, and green face. The back of the eyes and head are rimmed with a narrow yellow band. The black abdomen has blue markings on the sides with a dark stripe on the top. The colors are more solid than in the mosaic darners. Female and male heads are colored similarly, but the female abdomen tends to be darker and have purplish spots on the sides. The front vein (costa) of the wing of the female is yellow. The large size and green head and unmarked, green thorax stand out and help with identification in flight. In the hand, notice that this species lacks a T-spot and instead has a "bull's-eye" mark atop the head. Also notice the amazing colors of this dragonfly, which don't seem real.

This is the only darner that lays its eggs in the tandem position. Females lay eggs in decaying wood or submerged stems of aquatic plants. Pairs can often be seen flying around a pond perching on even the smallest of emergent plants to oviposit. This species has been recorded migrating southward along the Oregon coast in late September and October. It is well known to migrate and emerges in Florida and migrates to the Midwest of the United States and to southern Canada as early as March. The progeny of these early migrants emerge in the late summer and fall and begin moving south. The remarkable fact about this situation is that there is apparently a resident population whose larvae overwinter as well as the migrant population which passes through the larval stage in one season (Corbet, 1999).

The Common Green Darner can be found at ponds and other still waters, including some slow-moving streams up to 6,000 feet in elevation throughout Oregon. This species is established throughout the United States and southern Canada and up into Alaska. It has a remarkable world distribution, being known from such disparate places as Central America and China.

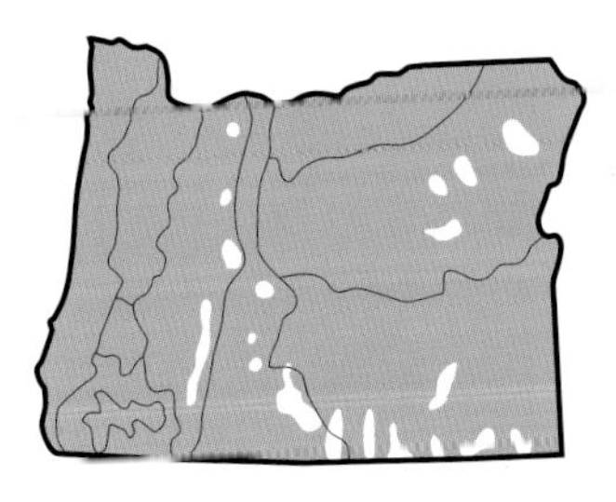

The adult flight season extends from late April to mid-November.

JAN	FEB	MAR	APR	MAY	JUN	JUL	AUG	SEP	OCT	NOV	DEC

Common Green Darner male

Common Green Darner female (inset: side view)

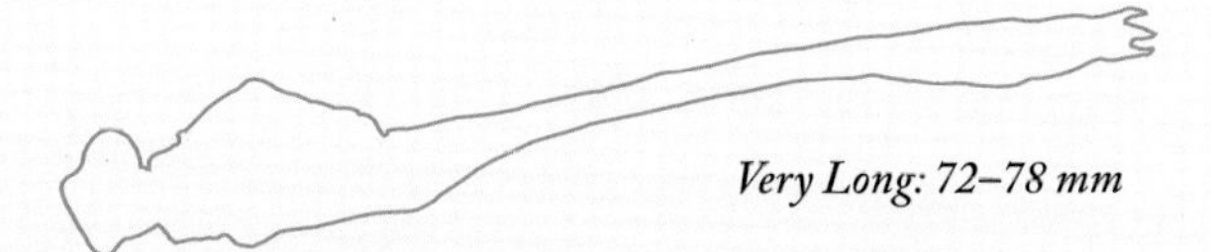

Very Long: 72–78 mm

Black-tipped Darner
Aeshna tuberculifera

Male Black-tipped Darners have beautiful sky-blue thoracic stripes which are rather straight and broad and may be green at the lower end. The front stripe on the sides of the thorax has a small extension to the rear at the top near the wing. The eyes are dull blue, and the face is green without a black line. There are blue spots at the rear on the top of the abdominal segments with the exception of segment 10 which is all black; thus, the name Black-tipped Darner. This is the only Oregon darner with simple appendages and no blue spot atop abdominal segment 10. The underside of the abdomen lacks spots. The male's terminal appendages are simple (i.e., not paddle-shaped) with a small tubercle on the underside of the upper appendages in side view. However, this feature will only be seen in the hand. Most females and males are colored similarly with green stripes on the sides of the thorax. Yellow-green colored females are also known. The cerci of the female are large compared to many of our other darners. The female has large cerci simulating the male, a thinner abdomen than other female darners, and has a flight behavior similar to males. Paulson (2009) postulates that simulation of the male enables the female to fly among males and oviposit with limited harassment.

The Black-tipped Darner is found from the east coast as far south as North Carolina into central Canada, the Midwest, and Washington on the west coast. However, it is absent in eastern Montana and the Dakotas. In 2009, it was first found in Union County, Oregon in a rush- and sedge-lined borrow pit at 4,100 feet in elevation. Here, males patrolled around and through the rushes from the tops of the rushes to near water level searching for females. Females oviposit from several feet above the water to just below the water surface in plant stems and floating vegetation.

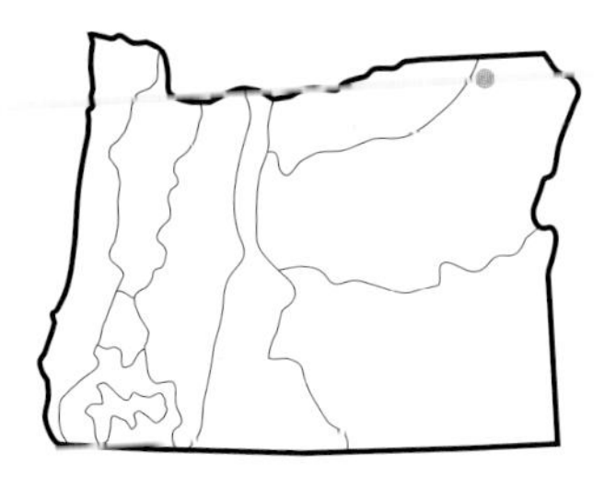

This species has been observed in early August, but the flight season in Oregon will obviously be extended with more observations.

JAN	FEB	MAR	APR	MAY	JUN	JUL	AUG	SEP	OCT	NOV	DEC
							■				

Black-tipped Darner male

Black-tipped Darner androchrome female • photo by Jim Johnson

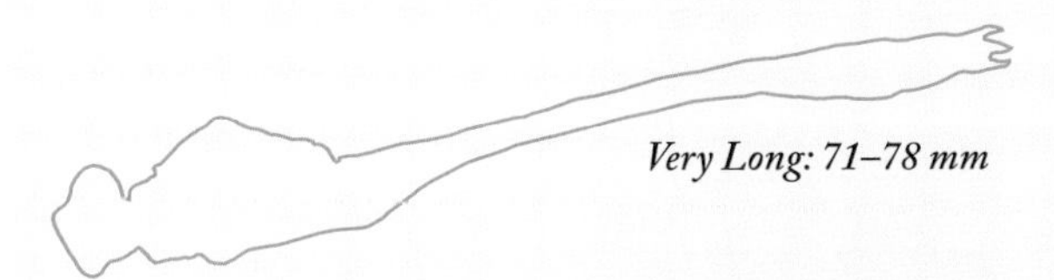

Canada Darner

Aeshna canadensis

The Canada Darner has blue stripes on the shoulder of the prothorax and rather broad blue stripes on the sides of the thorax. The top half of the front thoracic stripe is deeply indented at the front. The top of the abdomen is constricted at segment 3 and has paired blue spots at the rear of each segment, including the tenth segment, and on the underside of the middle abdominal segments. The face is pale blue without a strong stripe, and the eyes are a subdued blue color. The male's appendages are of the simple variety, gently curving up from front to rear like a small ski tip with small serrations on top of the upper appendage. Similar species include the Sedge, Shadow, and Subarctic Darners. When comparing this darner to these other species, note the color of the facial stripe, shape of the thoracic stripes, and shape of the cerci. There are both blue and green forms of the female. The green form has duller colored eyes and yellow-green rather than the blue male colors. The blue female and the male are very alike in color. The shape of the thoracic stripe is also characteristic for identification of the females.

This species is distributed across the northern United States and southern Canada. In Oregon, it is a Cascade Mountain species found above 3,000 feet in elevation. Males are often found patrolling the shoreline of boggy ponds and pools. Look for it at mountain lakes, bogs, streams, and wet prairies such as Gold Lake Bog, Big Marsh Creek, and Camas Prairie. At Gold Lake Bog, they prefer the smaller, shallow pools in the bog rather than the lake. Paulson (2009) notes that females oviposit in mosses and sedges at the edges of open water, and adults can live up to 70 days.

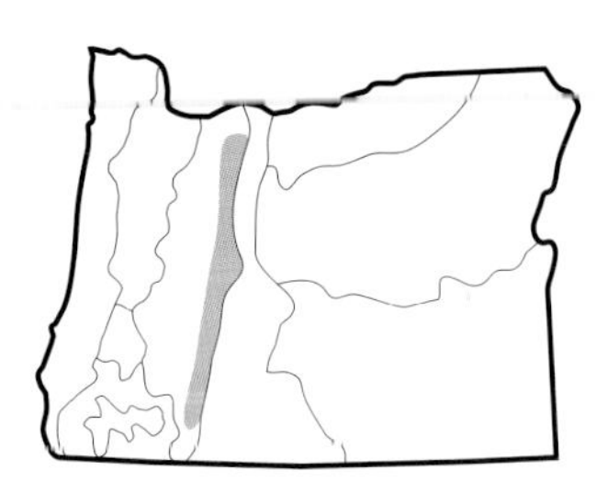

The adult flight season extends from late June to early October.

JAN	FEB	MAR	APR	MAY	JUN	JUL	AUG	SEP	OCT	NOV	DEC

Canada Darner male

Canada Darner female (inset: androchrome female)

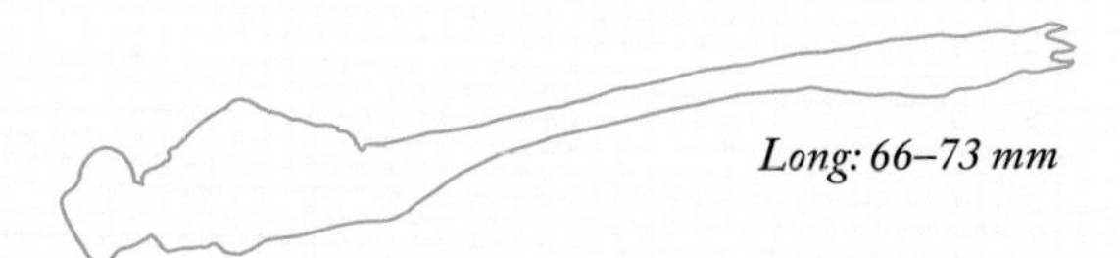

Long: 66–73 mm

Lance-tipped Darner

Aeshna constricta

The male Lance-tipped Darner has broad stripes on the sides of the thorax which are yellow to blue in color. The front stripe is indented at the front, and there is a small backward slash at the top. There are paired blue spots on the top of the abdomen, including the tenth abdominal segment, but no spots on the underside of the abdomen. The face is yellow-green without a black stripe. The male appendages are of the paddle type with a strong spine on the bottom of the paddle at the rear. This species is named after its large, lance-like appendages. The bottom of the paddle tends to be straighter and the spine thicker than in the Paddle-tailed Darner. Of the darners with a paddle-type "spiked" upper appendage, only the Lance-tipped and Paddle-tailed have a blue spot atop abdominal segment 10. The Paddle-tailed Darner has a facial stripe and the Lance-tipped Darner does not. Females and males may be colored alike, but some females may have lighter colored, green, thoracic stripes. Some females may be duller with yellow marking replacing the blue (Paulson, 2009) but we have not observed them in Oregon. Females have very large cerci which make them easy to separate from other darner species.

The Lance-tipped Darner is found across North America from central Canada as far south as New Mexico. Its range is very restricted in Oregon where it is currently known only from Lake of the Woods in Klamath County. It was discovered here by Jim Johnson during the 2005 annual Oregon Dragonfly Survey field trip. It was netted with a feeding swarm consisting of five species of darners (Gordon & Kerst, 2006). In British Columbia, the Lance-tipped Darner is found at small ponds and nutrient-rich marshes and may develop in seasonal waters (Cannings, 2002). Since this species is found in southern Washington as well as southern Oregon, the Lance-tipped Darner may occur at other Oregon locations.

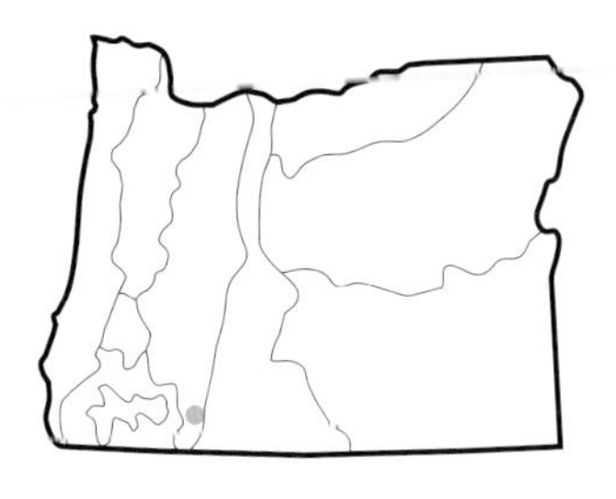

The adult flight season extends from early August to mid-September.

JAN	FEB	MAR	APR	MAY	JUN	JUL	AUG	SEP	OCT	NOV	DEC

Lance-tipped Darner male

Lance-tipped Darner androchrome female

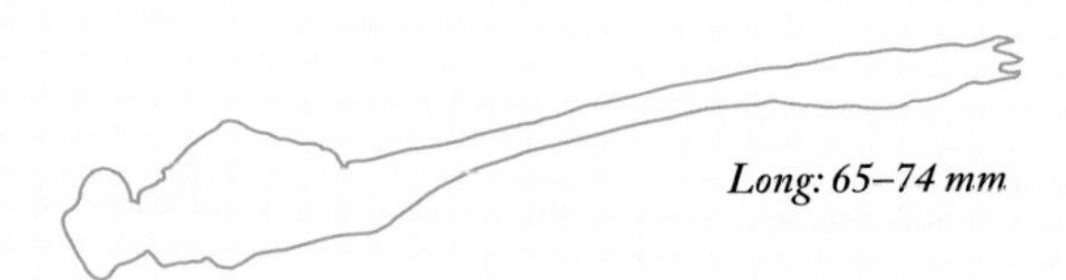

Paddle-tailed Darner

Aeshna palmata

This is one of our typical blue and black mosaic darners. The Paddle-tailed Darner has broad, slightly wavy, yellow-green to blue stripes on the sides of the brown thorax. The posterior stripe is broad at the top. The face is also yellow-green with a black line. The eyes are blue but much more subdued than the bright blue of the Blue-eyed Darner. Paired blue spots occur on the top of the abdominal segments, including segment 10. It does not have blue spots on the undersides of the abdomen. The male appendages are paddle-shaped, as the name suggests, with a trailing spine that is rather sharp. The female and male have similar color patterns, but both blue and yellow-green form females are known. The yellow-green females have brown eyes and a light brown pterostigma surrounded by black veins.

The Paddle-tailed Darner is a western species found from Alaska to southern California and as far east as Nebraska. It has been found flying after snowfall at over 4,000 feet elevation in southern British Columbia and is recorded from Colorado at 11,480 feet elevation (Corbet, 1999). It is one of our most abundant darners even in urban areas and is common throughout Oregon, frequenting ponds, lakes, and all types of still waters up to at least 7,500 feet in elevation. Males fly along the shoreline of water bodies searching for females—often hovering in one spot or venturing down into emergent vegetation. Also look for these darners feeding in open areas along trails, roadways and in campgrounds. Females oviposit in stems of emergent vegetation sometimes several feet above the water surface (Manolis, 2003).

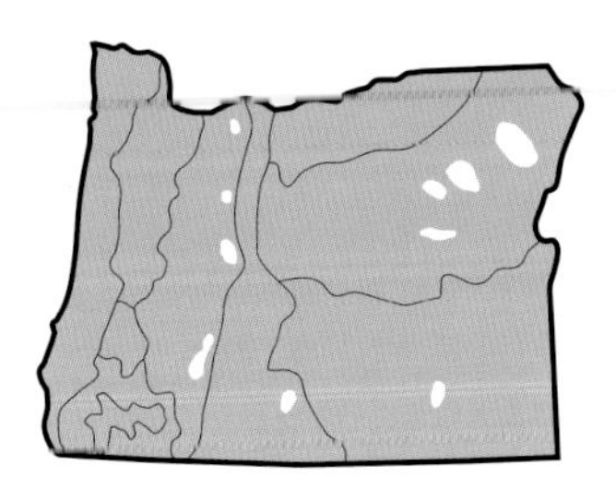

The adult flight season extends from late June to mid-November.

JAN	FEB	MAR	APR	MAY	JUN	JUL	AUG	SEP	OCT	NOV	DEC

Paddle-tailed Darner male

Paddle-tailed Darner female (inset: androchrome female)

Long: 64–77 mm

Sedge Darner

Aeshna juncea

The Sedge Darner is a very attractive species with broad yellow-green stripes bordered strongly in black on the sides of the thorax. The front stripe tapers towards the top while the rear stripe is narrow at the bottom and broadens at the top. The thoracic stripes are often yellow at the bottom transitioning to blue at the top. Paired blue spots occur on the top of the abdomen, including segment 10, and pale blue spots are also found on the bottom of the middle abdominal segments. The face is yellow-green with a black line, and the eyes are dull blue. The T-spot crossbar has a very blurred front edge. The male's appendages are of the simple variety. Females may be colored yellow-green or blue like the male.

In North America, the Sedge Darner is a northern species found across the United States and Canada to the Arctic Ocean. Cannings (2002) notes that it is the most common darner in the Yukon Territory of Canada. It occurs in the Rocky Mountains and as far south as the Cascades in Lane County. As the name implies, it frequents the sedge-lined shores of mountain lakes and wetlands from 2,500 to 7,500 feet in elevation. It was discovered at Gold Lake Bog in 2007 for the first Lane County record for this species. Larvae have been found alive under stones in ponds that have gone dry in British Columbia (Corbet, 1999). It is not a common species in most locations where it is found, with the exception of the Blue Mountains. We have found it to be very common at Anthony Lake, Grande Ronde Lake, and Mud Lake in the Elkhorn Mountains. Females oviposit in emergent vegetation at or below water level or in mosses (Paulson, 2009).

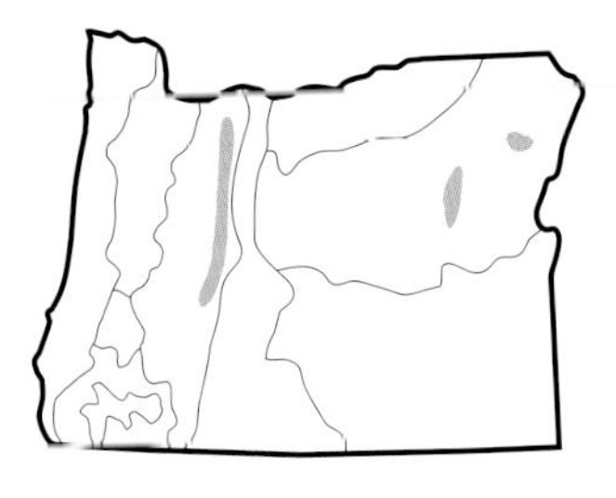

The adult flight season extends from mid-July to early October.

JAN	FEB	MAR	APR	MAY	JUN	JUL	AUG	SEP	OCT	NOV	DEC
						▬	▬	▬	▬		

Sedge Darner male

Sedge Darner androchrome female

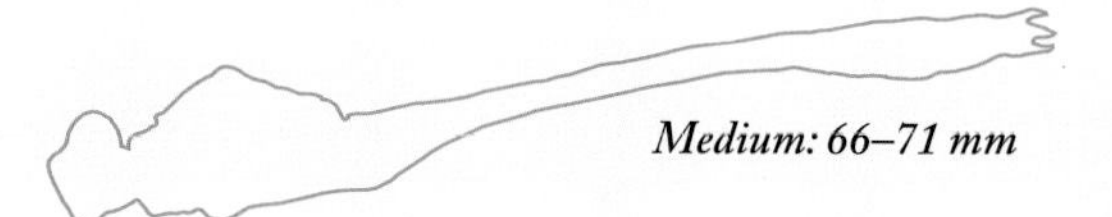

Medium: 66–71 mm

Shadow Darner

Aeshna umbrosa

This is another species with a paddle-shaped upper appendage with a trailing spike on the lower end. The male appendages are most similar to the Paddle-tailed Darner. The thoracic frontal stripes are green, and the relatively narrow, straight side stripes are yellow to green at the bottom and blue at the top. Each stripe has a rearward extension at its top. The abdominal spots are small compared to our other mosaic darners, and the top of segment 10 has no spots; the underside of the abdomen has paired blue spots on most segments, but they may be pale. The dull bluish face has no black stripe. Most Shadow Darners have a brown facial stripe but some individuals do have a thin black line. The male Paddle-tailed has a black facial stripe. The females are either colored yellow-green or blue as in the male.

The Shadow Darner is widely distributed across North America from Northern Canada to the southern United States. It is found throughout Oregon up to 6,000 feet in elevation and has been found in the Willamette Valley in urban areas. As its name implies, this common species often patrols shaded areas along the banks of streams, lakes, and ponds, including beaver ponds in Oregon. We have even seen it flying along the shaded banks of Wolf Creek in the Ochoco Mountains in the rain. Paulson (2009) observes that the Shadow Darner flies later in the day and may not be present earlier. Females oviposit in muddy banks or woody debris along the shoreline (Manolis, 2003). It is one of the last dragonflies to remain in the fall, often into late October and November depending on the weather. It has been seen flying in Oregon as late as mid-December.

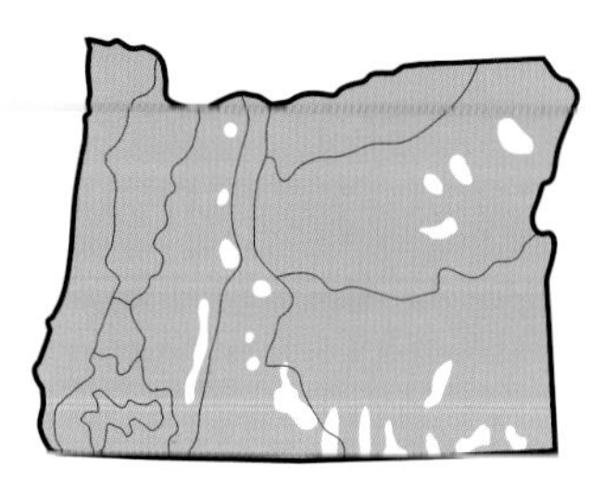

The adult flight season extends from late May to mid-December.

JAN	FEB	MAR	APR	MAY	JUN	JUL	AUG	SEP	OCT	NOV	DEC

Shadow Darner male

Shadow Darner male • photo by Jim Johnson

Long: 65–73 mm

Subarctic Darner

Aeshna subarctica

The male Subarctic Darner has thoracic stripes that are yellow-green at the bottom transitioning to blue at the top. This medium-sized darner has characteristic stripes that are rather narrow tapering to a constriction in the middle. The abdomen has fairly large paired blue spots on the top of the segments, including segment 10. The bottoms of the abdominal segments are without conspicuous spots. The face is yellow-green with a black line, and the eyes are blue. The T-spot stem is narrow and straight. The males' appendages are of the simple variety but are broad compared to other species with a slight outward flare at the tip. Females and males are patterned similarly. We have only seen blue females in Oregon, but in some females the blue is replaced with yellow or green spots (Paulson, 2009). The female has duller colored eyes and a yellow-green face with a black line.

The Subarctic Darner is another northern species. It is found across the northern United States, Canada, and Alaska. In Oregon, it is only known from Camas Prairie, Little Crater Lake, and Clackamas Lake in Wasco and Clackamas Counties at 3,000 to 3,500 feet in elevation. The Subarctic Darner is another species with a long pre-reproductive period and an adult life span that approaches 70 days (Corbet, 1999). While it is rare in Oregon, Camas Prairie seems to be excellent habitat for this species where it can be common at well-vegetated pools in the prairie during the flight season. In Washington, this species has only been found in two locations in the northeast, and it has not been found in California. Corbet (1999) observes that the Subarctic Darner is an example of a species with high-altitude relic populations of a boreal-alpine species that are separated from the main populations. This may explain the discontinuous distribution in the northwest.

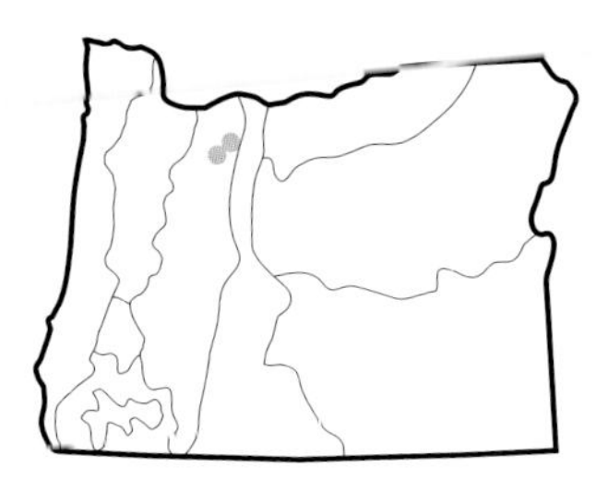

The adult flight season extends from late July to early October.

JAN	FEB	MAR	APR	MAY	JUN	JUL	AUG	SEP	OCT	NOV	DEC

Subarctic Darner male

Subarctic Darner androchrome female

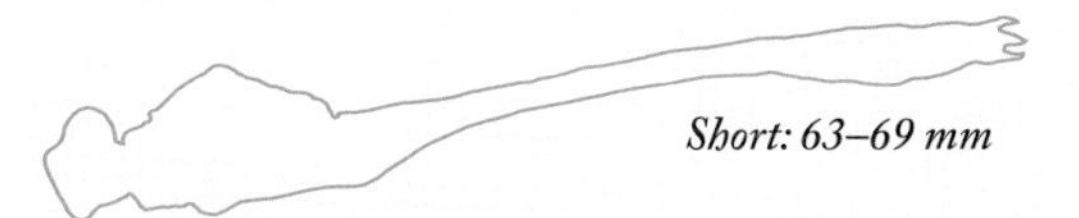

Short: 63–69 mm

Variable Darner

Aeshna interrupta

The Variable Darner is well named as the thoracic stripes on this species can vary from nonexistent to upper and lower partial stripes or a full stripe with a constriction in the middle. Both its scientific and English names reflect the features of the thoracic side stripes: variable and often interrupted. The thoracic stripes are blue as are paired spots on top of the abdomen, including segment 10. The underside of the abdomen does not have spots. The face is yellow-green with a black stripe. The T-spot cross-arm and stem are very thick, and the stem is very wide at the base. The males' appendages are of the simple variety. Females and males are similar in pattern, but the females can be either blue or yellow-green in color. With experience, this species can be identified in flight by the darker appearance of the thorax since it lacks the bright stripes of most other species. Several subspecies are recognized but are difficult to separate. Descriptions of these subspecies have been based on both thoracic striping and male appendages.

The Variable Darner is one of the common species throughout Oregon and is found across North America. It can be found at still waters mainly above 3,000 feet in elevation, is rare in the lowlands, and has been found at elevations near 10,000 feet in Utah (Corbet, 1999). It is found at mountain lakes, boggy ponds, and pools in meadows, and is one of the most often encountered darners in the mountains in Oregon. As with many of our other darners, males patrol near the shoreline of vegetation at ponds searching for females. Females oviposit in floating sedges and grasses, upright plant stems, and wet logs (Paulson, 2009). We have not found this species in the Willamette Valley.

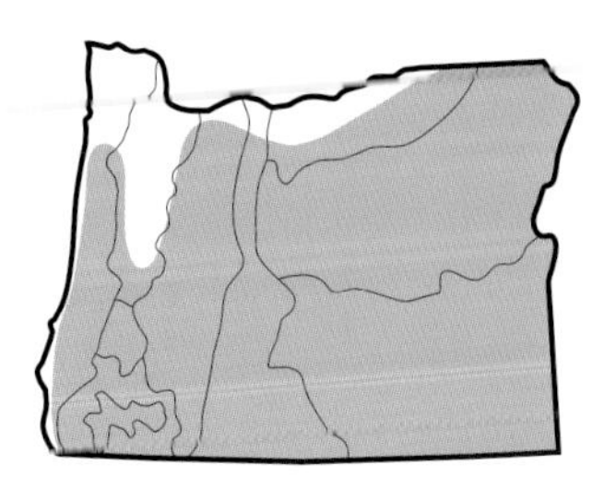

The adult flight season extends from late May to early October.

JAN	FEB	MAR	APR	MAY	JUN	JUL	AUG	SEP	OCT	NOV	DEC

Variable Darner male (inset: male with plain thorax)

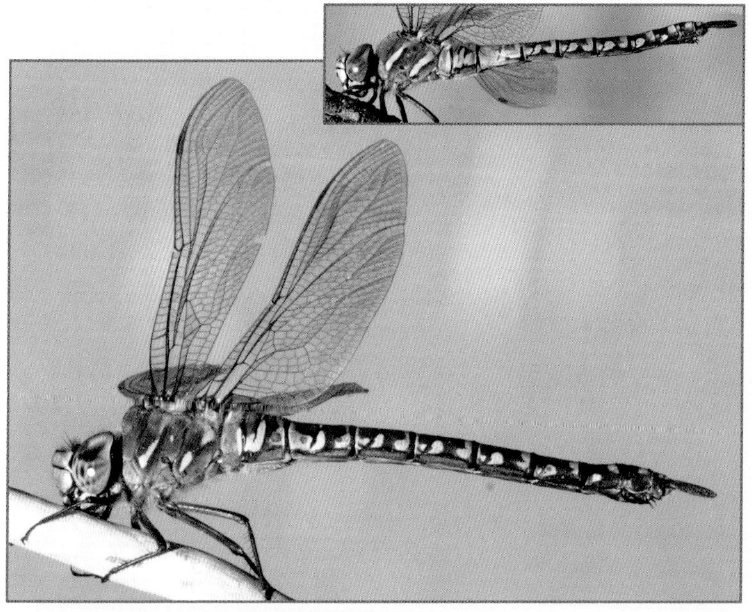

Variable Darner female (inset: androchrome female)

Medium: 67–71 mm

Walker's Darner

Aeshna walkeri

Walker's Darner was named by C. H. Kennedy after Edmund M. Walker, who was a renowned Canadian author and Odonatologist. This species is unique among our darners in having narrow, white or bluish-white thoracic stripes and a white face with a slight brown line. However, some individuals do show a black facial line. There are also large, paired blue spots on the top of the abdomen, but no spots on the top of abdominal segment 10 or the underside of the abdomen. The male appendage is paddle shaped and broad. The spine at the tip is thick and short compared to other darners with paddle-type appendages. In addition, the spine doesn't originate at the very tip of the paddle but is slightly forward. This set of characters generally makes this darner easy to identify. Females may have either blue or green abdominal spots. Care needs to be exercised during identification by color as some darner males with paddle-type appendages can have similar colors when teneral.

In Oregon, Walker's Darner has been found at only a few locations in north and south central Oregon including Gilliam, Sherman, Lake, Klamath, and Josephine Counties. However, it will likely be found at more locations in southern Oregon in the future. It is fairly common on Jenny Creek in Jackson County, a small, rocky stream bordering Highway 66 near the Pinehurst Inn, and this is the best spot in Oregon to see this interesting species. Males can be found patrolling low over water on streams where they search indentations along the shore for females. We have found females especially difficult to find. Females oviposit in mossy seams and submerged roots of plants (Manolis, 2003). We have also observed a female ovipositing in a vertical mud bank a foot above the water. This species is currently only known from California, where it is widespread, and Oregon, where its range is fairly restricted.

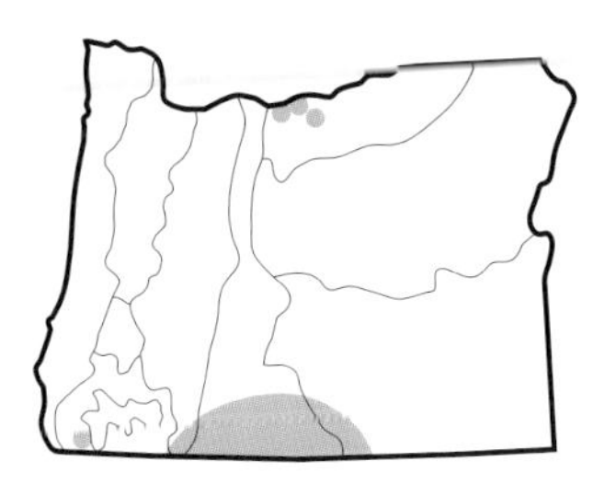

The adult flight season extends from mid-July to mid-October.

JAN	FEB	MAR	APR	MAY	JUN	JUL	AUG	SEP	OCT	NOV	DEC

Walker's Darner male

Walker's Darner androchrome female

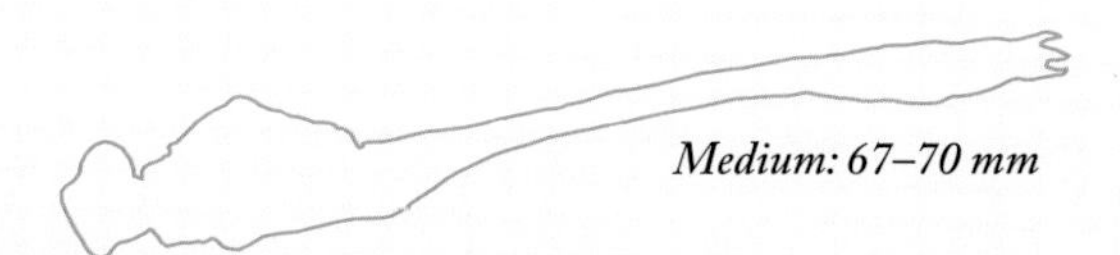

Zigzag Darner

Aeshna sitchensis

The common name of this species comes from the zig-zag shape of the narrow thoracic stripes, while the scientific name is derived from the type locality, Sitka, Alaska. The Zigzag Darner is small compared to most other darners. The thin thoracic stripes, especially the front stripe, are unique among our Oregon darners and rather like a lazy "Z"-shape. The color of the stripe is the usual blue to yellow-green from top to bottom. The top of the abdomen has paired blue spots, including segment 10. The abdomen is without markings on the underside and the male appendages are simple. The face is yellow-green with a black line. The base of the T-spot is unique, with the ends of the base convex upward toward the crossbar. The female also displays the same zig-zag thoracic stripes and may be marked with either blue or yellow-green.

The Zigzag Darner is found in the northern United States and across Canada into Alaska and at altitudes to 10,000 feet in Utah (Corbet, 1999). In Oregon, it is rare and has been found at only a few locations in Clackamas, Lane, and Deschutes Counties from 3,000 to 5,500 feet in elevation. These locations include wet, boggy spots like Gold Lake Bog and the meadow at Sparks Lake. At Sparks Lake, adults seemed to hang up in clumps of sedges in the meadow away from water. Unlike other darners, the Zigzag Darner often perches horizontally on the ground or on logs. Females oviposit in moss beds, algal mats, and mud at the edges of open water or into vegetation at water level (Paulson, 2009). This is another species that may survive dry periods in seasonal wetlands. A larva of the Zigzag Darner was found to have molted beneath a stone in a summer-dry pond (Corbet, 1999). Cannings (2002) notes that this species is restricted to peatland bogs and fens in British Columbia where the surface is covered by mosses and there is little open water. There are both seasonal and permanent pools in the meadow at both Sparks Lake and Gold Lake Bog.

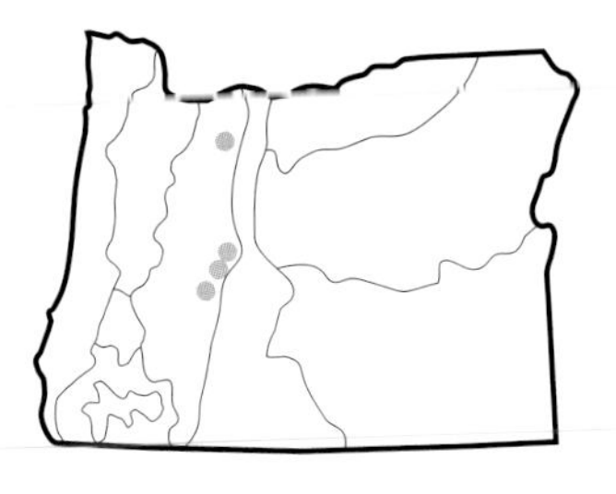

The adult flight season extends from late August to mid-September.

JAN	FEB	MAR	APR	MAY	JUN	JUL	AUG	SEP	OCT	NOV	DEC
							▬	▬			

Zigzag Darner male

Zigzag Darner androchrome female

Short: 55–68 mm

Blue-eyed Darner

Rhionaeschna multicolor

This is a truly spectacular, blue-patterned dragonfly with bright blue eyes and a face which lacks a black facial line. The brown facial line can appear fairly strong. These head features, particularly the bright, blue eyes and blue face can be seen while in flight, especially when it is hovering. In the hand, it is our only darner with forked upper appendages. The thoracic frontal stripes are blue and the side stripes are pale blue, straight, and fairly broad. There are paired blue spots on the top of the abdomen including segment 10. The underside of the abdomen is without spots. Females may be either blue as on the males or yellow-green. The face and eyes of the female are brownish-green, and the thorax has fairly broad green stripes. The wings of the females have a brown tint. This genus represents a Neotropical group which exhibits a tubercle under abdominal segment 1 (see Figure 3, page 74).

The Blue-eyed Darner occurs from the Midwest to the west coast in the United States and into southern British Columbia in Canada. It is one of our common darners even in urban areas throughout Oregon at ponds, lakes, and at slow-moving waters. While it occurs up to 6,000 feet in elevation, it is less common above 4,000 feet. Females oviposit above or below the waterline in emergent vegetation, floating plant stems, and woody branches in open water (Paulson, 2009). Watch for it patrolling a few feet above the water, where it will hover at times, bright blue eyes showing. We have watched them flying until dark on the Owyhee River at Rome—coming in to hang up in trees eight to ten feet above the ground when it was so dark that they were only visible against the sky.

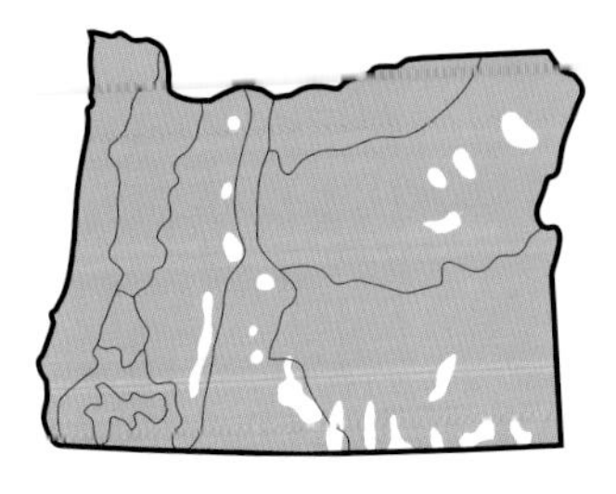

The adult flight season extends from late May to late October.

JAN	FEB	MAR	APR	MAY	JUN	JUL	AUG	SEP	OCT	NOV	DEC

Blue-eyed Darner male

Blue-eyed Darner female

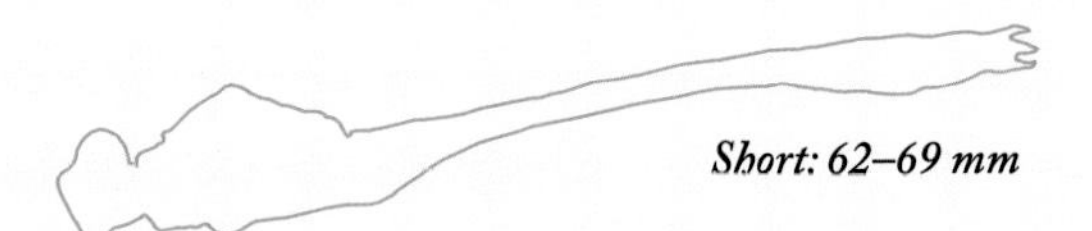

Short: 62–69 mm

California Darner
Rhionaeschna californica

This darner is the smallest of our Oregon darners, and its smaller size is noticeable in flight. With experience, it can be separated from our other darners, even in flight, because of its small size. The pale blue thoracic side stripes are straight and narrow and bordered in black. The front side stripe on the thorax is narrowed at the top. There are small spots on the front of the thorax and paired blue spots on the top of the abdomen, including segment 10. The underside of the abdomen does not have blue spots. The pterostigma is brown. The California Darner has simple, rounded appendages with no trailing spike. The eyes are blue above and brown below, and the face is bluish-white with a black line. The T-spot atop the head has a very thick stem. As with most other darners, females and males may be colored alike, or the females may have yellow-green markings. The yellow-green female has brown eyes, a yellow-green face with a black line, and light brown pterostigma. Like the Blue-eyed Darner, this species is a member of the genus, *Rhionaeschna,* a Neotropical group which exhibits a tubercle under abdominal segment 1 (see Figure 3, page 74).

Although named for its type locality at Mt. Tamalpais, California, this species occurs across the western United States and into southwestern Canada. This common dragonfly frequents ponds, lakes, and slow-moving streams all across Oregon up to 7,300 feet in elevation. It is found from the Willamette Valley and Cascade Mountains to the alkali lakes of eastern Oregon. Larvae of the California Darner have even been found alive in water bodies that have gone dry in California (Corbet, 1999). Males patrol close to the water along the shoreline searching for females. Females oviposit similar to other darners in vegetation at water level.

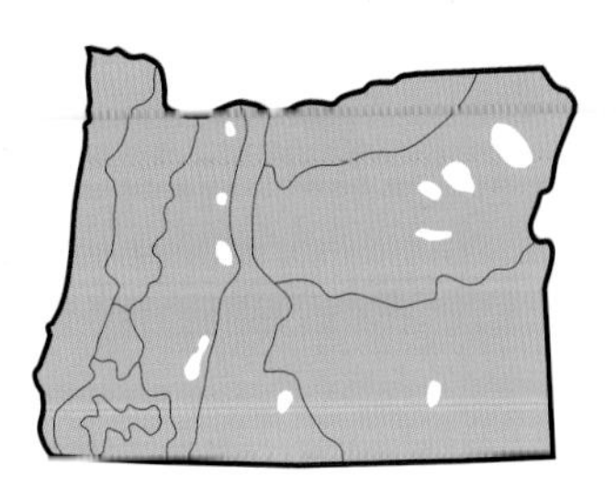

The California Darner has been found flying as early as mid-April in Oregon. It is one of our earliest darners to fly in the spring but can be hard to find later in the season.

JAN	FEB	MAR	APR	MAY	JUN	JUL	AUG	SEP	OCT	NOV	DEC

California Darner male

California Darner female

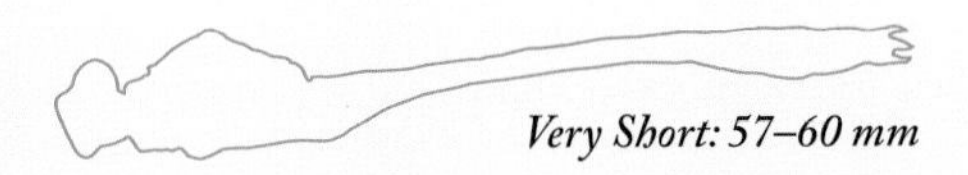

Clubtails, Family Gomphidae

The common name of the clubtails refers to the expanded tip of the abdomen, although this clubbed tip is more evident in males than in females. The clubtails are a favorite of many dragonfly enthusiasts due to their size, color patterns, and interesting behaviors. The clubtails are medium-sized dragonflies that spend a lot of time perching. Favorite perch spots are located in sunny areas on rocks, the ground, and vegetation of low to medium height near water. At times they are found away from water, especially in the sagebrush country of eastern Oregon. Clubtails have the eyes widely separated on the head. The short legs are well adapted to the perching behavior. Males will often fly up from a perch out over open water while defending a territory or looking for females and back to perch again. They can be identified in flight by their low undulating and zig-zag pattern of flight, often flying just above the water surface.

After mating, females oviposit unguarded by tapping the water surface with the tip of the abdomen releasing many eggs each time. They may be seen flying down from streamside trees to tap the water and flying back up to perch.

Clubtail larva (Snaketail)

Clubtail larvae are burrowers found in streams and rivers with the exception of the Pacific Clubtail (*Gomphus kurilis*), which is sometimes found at ponds and lakes. In other areas of the country, pond and lake clubtails are more prevalent. The larvae are more flattened and elongate than some other dragonfly larvae, an adaptation to their burrowing behavior. Burrowing clubtail larvae are able to bury themselves rapidly, move around below the substrate surface, and locate prey using mechanoreceptors (Corbet, 1999). Clubtail larvae crawl up on rocks and banks when ready to emerge, rather than vegetation, as the legs are not well adapted to climbing (Needham, et. al., 2000).

Oregon hosts nine of the 100 North American clubtail species. Identification of these nine species can be simplified by remembering that the abdominal clubbed tip is a key feature and is more evident in males. Females show a much more cylindrical abdomen. With minor exceptions, males and females are patterned and colored similarly. As clubtails age, they often transition from yellow to green to dull green.

The nine Oregon species of clubtails include a single species in the genus *Erpetogomphus*, the ringtails. This dragonfly genus, found only in eastern Oregon, is one of our most strikingly colored with its black and white abdomen. Oregon has two species in the *Gomphus* genus both having rather restricted distributions in Oregon. The single species in the genus *Octogomphus* in Oregon is restricted to western Oregon and can be quite common at some locations. Where it is common, this species (Grappletail) is fun to watch for its interesting and dizzying behavior. The four species of *Ophiogomphus* in Oregon are known as snaketails and one species or another is found throughout the state. They are quite attention-getting with their patterned green thorax and black and yellow abdomens. Our single species in the genus *Stylurus*, the hanging clubtails, has a very restricted range in Oregon. Although Oregon does not have a large number of species in this family, the fauna is quite diverse with five genera represented. The Oregon species of Gomphidae can be organized into four groups for identification purposes based on the abdominal patterns. Two charts (see Identification Charts H and I) illustrate key features which help identify these four groups.

Sinuous Snaketail male

Identification Chart H • Clubtails, Family Gomphidae, Genera: *Erpetogomphus, Octogomphus, Gomphus*, and *Stylurus*

Species (size)	Abdomen (top view)	Thorax (side view)
White-belted Ringtail ***Erpetogomphus compositus*** Medium 46–55 mm	 (side view) Black and white rings  (top view) Black and white rings; orangish S7–10, unique	White side bands with yellow and dark bands, diagnostic
Grappletail ***Octogomphus specularis*** Short 51–53 mm	 Dark with thin, yellow median line; S10 enlarged with grappling hook appendages; distinctive	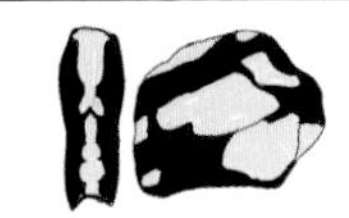 top: Black with yellow "goblet" mark; side: Yellow surrounded by black

Genera *Gomphus* and *Stylurus*: Abdomen mostly dark with thin, yellow, arrow-shaped markings pointed rearward; appendages dark

***Gomphus* and *Stylurus* Species (size)**	Abdomen (top view)	Thorax (side view)
Columbia ***Gomphus lynnae*** Long 54–60 mm	 Thin, dart-shaped, yellow markings pointed rearward; yellow spot atop S10; deeply forked appendages	 Broad dark shoulder; two side stripes; the second behind the hindwing base; shows pruinosity with age
Pacific ***G. kurilis*** Medium 52–56 mm	 Thin, needle-shaped, yellow markings pointed rearward; S10 black on top	 Single side stripe
Olive ***Stylurus olivaceus*** Long 55–60 mm	 Thin, dart-shaped, yellow markings pointed rearward; S10 black on top	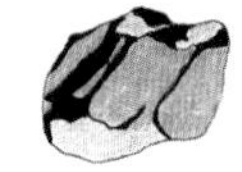 Olive; no side stripe

Snaketails, genus *Ophiogomphus*: Abdomen with more yellow and less black than genera *Gomphus* and *Stylurus*; appendages yellow (see next chart)

Identification Chart I • Snaketails, Genus *Ophiogomphus*

Snaketails, genus *Ophiogomphus*: Abdomen with more yellow and less black than *Gomphus* or *Stylurus* Clubtails in previous chart; appendages yellow

Snaketail Species (size)	♂ Appendages (side view)	Abdomen (top view)	Thorax Shoulder Stripe (side view)
Bison *Ophiogomphus bison* Short 49–51 mm	upper; Long, straight; lower: Long, upturned hook at tip	 S10 mostly yellow with black on basal edges	Broad, dark brown
Pale *O. severus* Short 48–52 mm	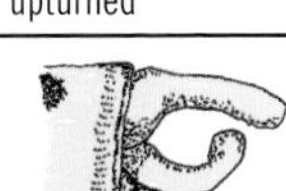 upper: Very stout; larger than lower; lower: Stout, upturned	 Broad yellow arrows pointed rearward; S10 all yellow	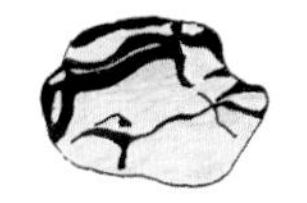 Variable: typical oblong spot or lacking spot
Sinuous *O. occidentis* Short 50–52 mm	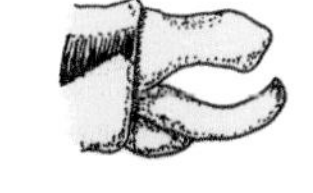 upper: Short, stout, longer than lower; lower: Sharply curved upward (like a coat hook)	Yellow arrows pointed rearward; S10 all yellow	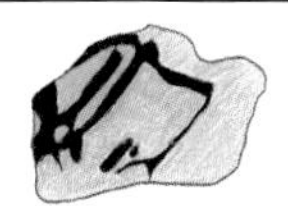 Wavy brown, thin; wavy pale stripe between
Great Basin *O. morrisoni* Short 50–53 mm	upper: Short, stout; lower: Longer than upper, gently curved upward	S10 black and yellow; yellow dots on top	Dark brown with thin pale line (less wavy than Sinuous)

Figure 4 • Female Snaketail Occipital Horns

(top view of head)

Bison Snaketail
Ophiogomphus bison

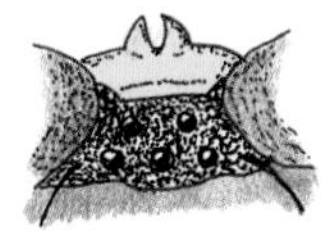

Horns closely spaced, "bison-like"

Sinuous Snaketail
Ophiogomphus occidentis

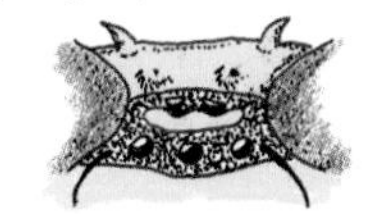

Horns spaced far apart

White-belted Ringtail

Erpetogomphus compositus

This species is one of six North American ringtails, and it is so striking in appearance that you can easily identify it when seen perched and in flight. The thorax is marked with stripes of yellow-green, dark brown and white (the only clubtail with white thoracic markings, which are the source of its English name "White-belted"). The base of the abdomen shows extensive white, and the middle abdominal segments are marked dramatically with alternating black and white rings and the terminal segments at the clubbed tip are orangish-yellow. This is our only clubtail with such a distinctively colored abdomen. The face is off-white, and the eyes are aqua-gray. It has short legs with pale femora and black tibiae. The wings are often tinged with amber at the base and have a black pterostigma. Males and females are similarly colored.

The White-belted Ringtail's range extends from southeastern Washington southward in the arid west into Mexico. In Oregon it is found in eastern Oregon up to 4,500-foot elevations and is most common in the John Day, Malheur, and Owyhee river basins.

The White-belted Ringtail is associated with streams and rivers. Here males frequently perch on rocks in the stream or on the shore. Males and females may also perch atop low bushes and sagebrush proximate to the streams. One can often see females in the sagebrush with the abdomen tipped up into the air like a flag.

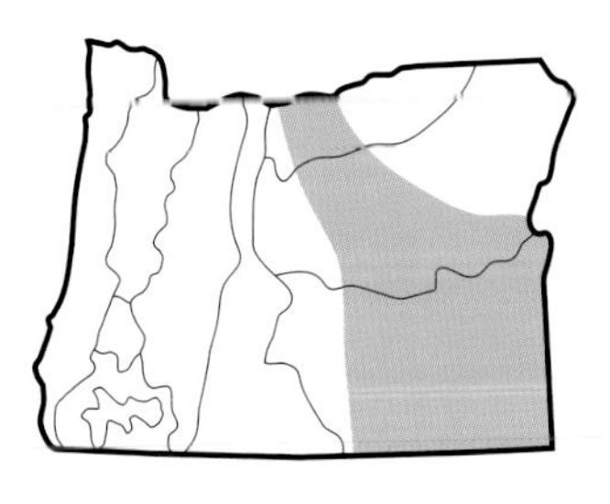

The adult flight season extends from mid-June to mid-September.

JAN	FEB	MAR	APR	MAY	JUN	JUL	AUG	SEP	OCT	NOV	DEC

White-belted Ringtail male

White-belted Ringtail female

Medium: 46–55 mm

Pacific Clubtail

Gomphus kurilis

This is a dark clubtail with a thorax that is yellow-green and dark brown to black. The top of the thorax has a dark middle stripe, the wide dark shoulder stripe has a thin yellow-green stripe in the middle, and the side of the thorax has a dark stripe that is narrower than the shoulder stripe. The abdomen is black and the top of the segments show long tapered, spearhead-shaped, yellow spots pointed toward the tip, with yellow spots on segment sides. The yellow spots on the side of the middle abdominal segments are small. The size and presence of spots atop segments 8 and 9 can vary but segment 10 is always black on top. The female and male are colored similarly. The face is yellow and the eyes are bluish-gray. The legs are black. The bright yellow fades in older specimens.

The Pacific Clubtail's range is strictly western and extends from Washington (where it is uncommon) south into central California. In Oregon it is most common in southwestern Oregon, from Lane County south, including coastal lakes.

The Pacific Clubtail is interesting in its ability to occupy varied types of habitat. It occurs from rock bottomed lakes, to large rivers, sand-bottom lakes, muddy ponds, and slow streams from sea level to almost 5,000 feet in elevation. It has been recorded in Linn and Benton Counties, and there is a record from Little Cultus Lake in Deschutes County. When stream pools dry up, larvae have been known to crawl across dried mud to reach another wet pool (Needham, et. al., 2000).

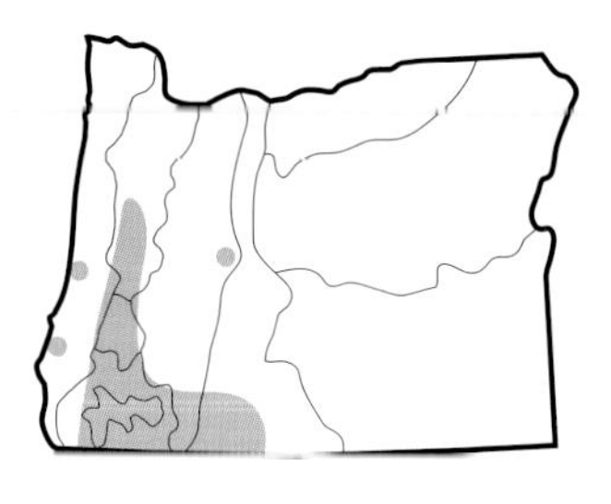

The adult flight season is from mid-May to early September.

JAN	FEB	MAR	APR	MAY	JUN	JUL	AUG	SEP	OCT	NOV	DEC
				▬	▬	▬	▬	▬			

Pacific Clubtail male

Pacific Clubtail female

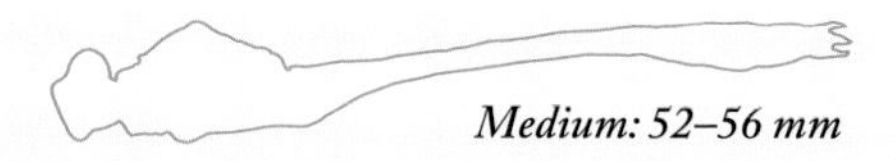

Medium: 52–56 mm

Columbia Clubtail
Gomphus lynnae

The Columbia Clubtail's thorax is yellow-green with bold, dark shoulder and side stripes and a second side stripe behind the base of the hindwing. The dark shoulder and middle side stripes are broken with yellow-green streaks. This species becomes pruinose on the head, thorax, abdomen, and upper legs as it ages, especially on the dark thoracic stripes and first two abdominal segments, making it unique among Oregon clubtails. Both sexes, which are colored similarly, show this pruinosity, which is also unusual. The top of the abdomen is dark with bright pale markings at the base and tip and pale yellow arrow-shaped markings in the middle sections, which taper in size toward the tip. The sides of the abdomen show small yellow and pale arrowhead markings that match the color of markings atop the abdomen. The face is yellow, and the eyes are aqua-blue. The legs are dark brown and the femora can show pruinosity. The front wing edge is yellow and the pterostigma is brown. In side view, the male upper appendages are nearly straight and show a point on the underside about half-way along the length. In top view, the upper appendages are divergent, and the lobes on the lower appendage are even more widely divergent.

The Columbia Clubtail's range is limited to Oregon and Washington in the Yakima, John Day, and Owyhee River basins. Dennis Paulson (1983) described this species, which he first discovered in Washington in 1971. It is the most recently described species of Odonata found in Oregon. It was first recorded in Oregon by Duncan Cuyler at the John Day River in Wheeler County on July 19, 1993 (Valley, 1993) during an annual meeting of the Dragonfly Society of the Americas.

Columbia Clubtails can be found along rocky shores and gravel bars along the John Day and Owyhee Rivers below the 3,500-foot elevation. They perch on rocks along and in the stream.

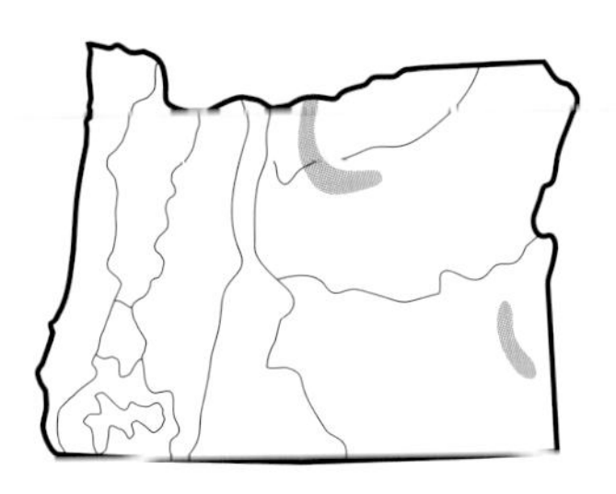

The known Oregon adult flight season extends from mid-June to mid August.

JAN	FEB	MAR	APR	MAY	JUN	JUL	AUG	SEP	OCT	NOV	DEC

Columbia Clubtail male

Columbia Clubtail female

Long: 54–60 mm

Great Basin Snaketail female

Olive Clubtail eating grasshopper

Grappletail

Octogomphus specularis

This is a unique Clubtail species and the only species in the genus *Octogomphus*. The Grappletail does not exhibit the typical flared, clubbed tail. Instead, it has an enlarged tenth abdominal segment with dynamic, forked upper appendages that resemble a grappling hook, and this is the origin of its English name. The genus prefix "octo" derives from the eight forked barbs, two each on the upper appendages and four on the lower appendage. The thorax is black above with the wide, black shoulder stripes setting off a middle yellow mark that resembles a goblet (see Figure 5, page 116). A wide yellow mark extends between the wings onto the tops of the basal abdominal segments. The sides of the thorax are yellow with a thin black line between the second and third thoracic segments. The abdomen is mostly dark above, but a thin yellow median streak extends from segments three to nine, and the swollen segment ten has a yellow spot atop it. The wide, black, upper appendages have extensive yellow markings on top. The female and male thoraxes are marked similarly, and the side of the female's abdomen shows a series of yellow spots and dashes. The face is yellowish with a black crossline. The widely separated eyes are gray, and the yellowish-green face has a black line through it. The pterostigma is thick and the hindwing is very large. The legs are black. As with several other clubtails, young adults are bright yellow but turn green and to dull gray-green with age.

Grappletails are very easy to approach. When you have a Grappletail in hand, be sure to examine the unique appendages, including the four-branched lower appendage. On the female, examine the four ridges on top of the head behind the eyes. Also look at the wing veins through your hand lens to see the many tiny barbs.

Grappletails favor swift streams and creeks in the mountains below 5,000 feet in elevation but can be found in small streams, seeps, fens and other slower waters.

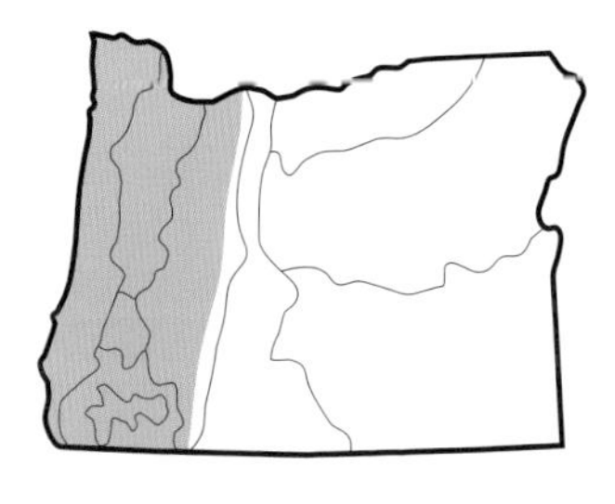

The Oregon adult flight season is from late May to early September.

JAN FEB MAR APR MAY JUN JUL AUG SEP OCT NOV DEC

to higher elevations to breed (Manolis, 2003). Females oviposit in riffles and the larvae live among leafy matter (Needham, et. al., 2000). Look for males on rocks or logs in or near the stream and for females in brush and small trees near the stream.

The Grappletail is a far western species with a range extending from British Columbia through southern California and into northern Mexico. In Oregon, its range extends from the Cascade crest westward with some local populations on the east Cascade slopes. In Lane County it has been found along the Coast Fork near Cottage Grove and at the Amazon Creek headwaters in Eugene. The bridge over Salt Creek at the Gold Lake outlet has an exceptionally high population density of Grappletails (see site 17 in the Dragonflying Spots section).

Figure 5 • Grappletail Features
***Octogomphus specularis* (top views)**

S10 with yellow dot, hooked appendages

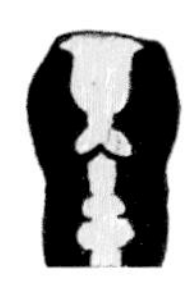

Thorax with distinctive "urn"-shaped yellow mark

The Wonder of Dragonflies

Grappletail male

Grappletail male (bright colors turn gray with age)

Grappletail female

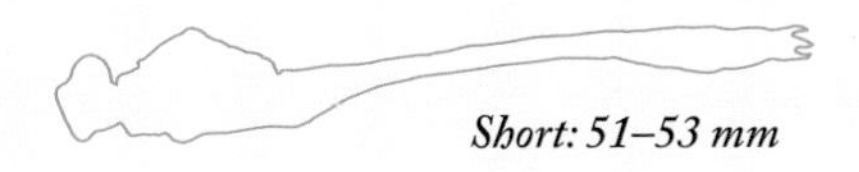

Short: 51–53 mm

Bison Snaketail

Ophiogomphus bison

Early emerging Grappletails can be found at lower elevations, and then they fly The thorax is bright green with broad dark brown shoulder stripes fused into a broad, straight band and a dark stripe on the shoulder of the thorax. The thoracic side stripe is a narrow line. The abdomen is black with dull yellow, triangular markings on top in a typical snaketail pattern. There are small, pale markings on the sides of the middle abdominal segments. Segment 10 and the top of the upper appendages are dull yellow. The face is green, and the eyes are gray. The bottom of the femur is pale, and the tibiae are black.

On the top of the female's head are a pair of small horns that are spaced closely together. These horns are shaped like a bison's horns, thus the name (see Figure 4, page 107). In the hand, in side view, the male's upper appendages are long and straight. The lower appendage is almost as long as the upper appendage and has a gently upward-curving point at the tip (see Identification Chart I).

Males can be found perched on gravel bars and bedrock. They fly out low over the water in rapid, zig-zag fashion. Look for females perched in grass or low brush away from the shore (Manolis, 2003). We have observed the females perching up in trees overhanging the water from which they fly down to oviposit in riffles and return to the trees. Males patrol riffles, awaiting the females.

The Bison Snaketail is a western species (Oregon, California, Nevada and Utah) with an Oregon range restricted to southwestern Oregon in the Rogue and Umpqua River basins. It can be found along lowland and foothill streams and rivers from 500–3,400 feet in elevation.

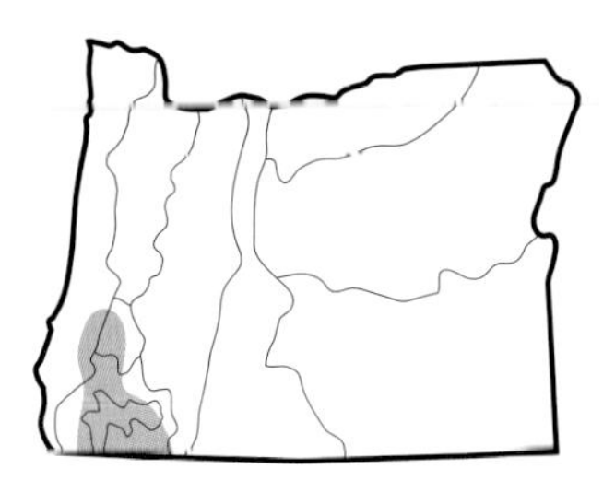

The adult flight season is short, extending from early June to early July.

JAN	FEB	MAR	APR	MAY	JUN	JUL	AUG	SEP	OCT	NOV	DEC

Bison Snaketail male

Bison Snaketail female

Short: 49–51 mm

Great Basin Snaketail

Ophiogomphus morrisoni

The thorax varies from gray-green to a pale lime green with a straight, fairly narrow, black or brown shoulder line separated by a thin, relatively straight pale green line. This line is not as wavy as on the Sinuous Snaketail. This character becomes clear with experience and makes the Great Basin and Sinuous Snaketails easier to separate. The Great Basin Snaketail displays a typical black and yellow snaketail abdominal pattern. The face is yellow-green with a narrow black crossline. The eyes are bright blue to bluish-gray. The femur shows a pale stripe on the bottom and the tibiae are pale on top. The front wing edge is yellow and the pterostigma is brown.

In the hand, in side view, the male's yellow upper appendages are short and stout and the yellow lower appendage curves upward but is longer than the upper appendage. Typically, the female's occiput has no horns, but on occasion may contain horns.

The Great Basin Snaketail's range is confined to Oregon, California and Nevada. In Oregon, its range is fairly limited to sites east of the Cascade crest, in the Klamath Basin, and in southern Oregon, but it was found at Salt Creek at the outlet to Gold Lake in 2004. In California, it has been found above 10,000 feet in the southern Sierra Nevada Mountains (Manolis, 2003).

The Great Basin Snaketail is usually found on rocks, logs, and gravel bars along small, cold, mountain streams and in the warmer, slower Sprague River in Klamath County—a river which receives considerable agricultural runoff. It is especially common at Crescent Creek in Klamath County.

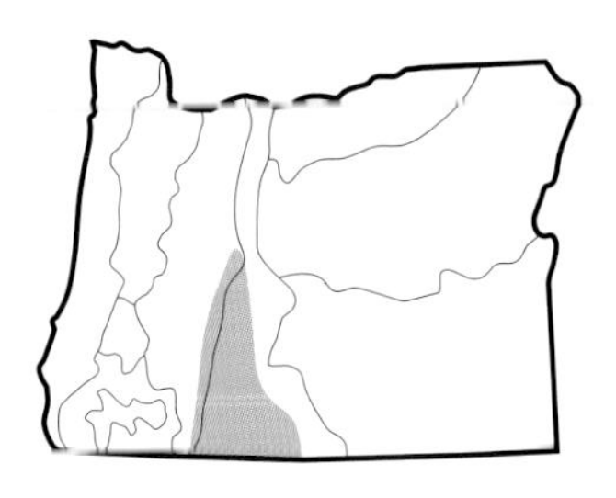

The adult flight season lasts from late May to late September.

JAN	FEB	MAR	APR	MAY	JUN	JUL	AUG	SEP	OCT	NOV	DEC

Great Basin Snaketail male

Great Basin Snaketail female

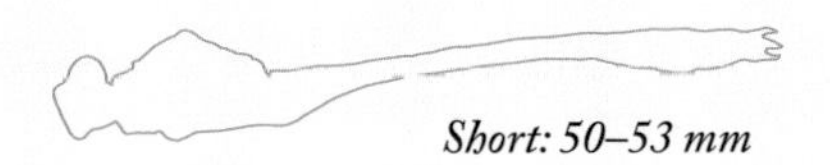

Short: 50–53 mm

Sinuous Snaketail

Ophiogomphus occidentis

The Sinuous Snaketail has a pale green thorax with wavy, dark brown shoulder stripes with a thin, sinuous pale stripe separating the two brown stripes—a key feature that results in its name. It shows a dark center stripe on top of the prothorax with a thin, partial center stripe. It may show a black stripe on the side behind the hindwing base. Like the other snaketails, its upper abdomen is patterned with black and yellow markings with segment 10 yellow. The eyes are dark gray and the face is greenish-yellow and plain. The femora are pale and the tibiae are dark. Males and females are similar.

In the hand, in side view, the male's yellow upper appendages are longer than the lower appendage, which is curved sharply upward and in side view resembles a wall-mounted coat hook. The female has occipital horns that are widely separated (see Figure 4, page 107).

The Sinuous Snaketail is a western species with a range extending from southern British Columbia into northern California and ranging as far east as Montana and Utah. In Oregon it is found east of the Cascade Mountains and in southwestern Oregon. In the Willamette Valley, it is found in Lane County and less commonly north to Marion County. It has been found along streams and in gardens in the Eugene urban area, likely traveling from the Willamette River.

It often frequents sandy and stony shores and nearby exposed sunny areas near a wide range of stream and river types up to a 4,300-foot elevation.

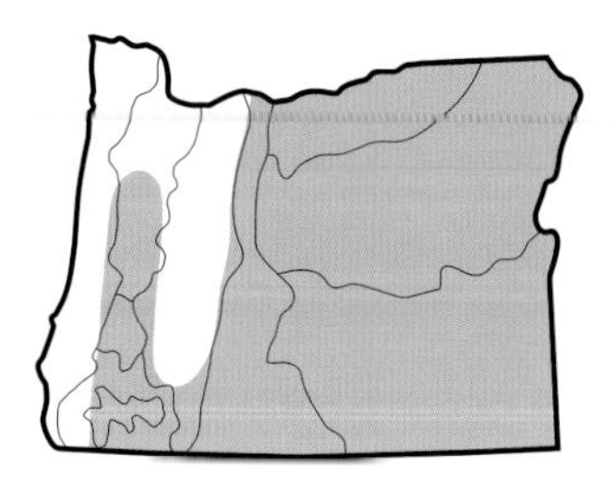

The adult flight season lasts from early June to mid-August.

JAN	FEB	MAR	APR	MAY	JUN	JUL	AUG	SEP	OCT	NOV	DEC

Sinuous Snaketail male

Sinuous Snaketail female

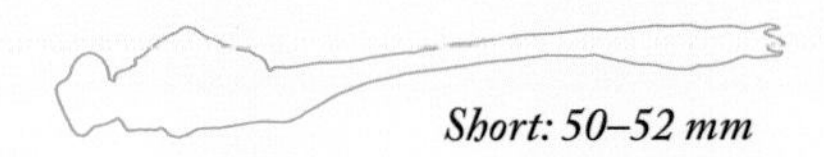

Short: 50–52 mm

Pale Snaketail

Ophiogomphus severus

The Pale Snaketail's olive-green thorax shows a fine brown middle stripe on top and a dark oblong or oval spot on the shoulder of the thorax. This shoulder spot can be missing on some individuals. There is also a thin dark stripe on the side of the thorax. The snaketail-patterned abdomen is dark brown to black with mostly yellow or whitish marks on the lower sides and large, yellow spots on top. The face is yellow and the eyes are aqua blue. The front wing edge is yellow and the pterostigma is grayish-brown. Females show more yellow on the legs than do the males.

In the hand in side view, the male's yellow appendages include a stout upper appendage and the lower appendage is about as long as the upper one.

The Pale Snaketail prefers clear, small, rocky streams and rivers, but has been found along muddier waters. It perches on the ground on streamside rocks, on emergent rocks in streams and on top of low bushes. The nymphs sometimes emerge on floating water lily leaves (Cannings, 2002).

The Pale Snaketail's range extends from British Columbia south into northern California and east into Nebraska. In Oregon its range includes eastern Oregon and the interior valleys as far north as the southern Willamette Valley. It has been found at elevations up to 6,000 feet. In Colorado, Pale Snaketails have been recorded above 9,000 feet in elevation (Corbet, 1999).

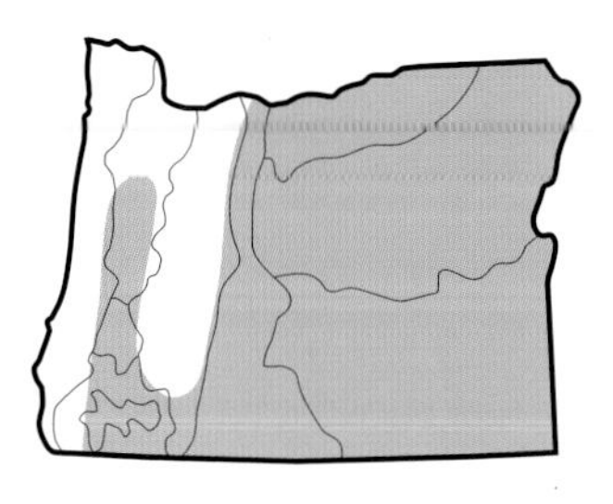

The adult flight season lasts from mid-May to mid-August.

JAN	FEB	MAR	APR	MAY	JUN	JUL	AUG	SEP	OCT	NOV	DEC

Pale Snaketail male (inset: male with unmarked prothorax)

Pale Snaketail female

Short: 48–52 mm

Olive Clubtail

Stylurus olivaceus

This species is in the Genus *Stylurus*, also known as the "hanging clubtails" because they tend to hang in vegetation, unlike other members of this family. The Olive Clubtail gets its names from its overall olive-green coloration.

The large Olive Clubtail has a pale olive-green thorax with a partly broken dark brown shoulder stripe and only a very thin dark line between the second and third thoracic segments. The top of the thorax has a light-colored triangular mark in front of the forewing and an olive-green, inverted "U"-shaped mark at the front of the thorax. The basal abdominal segments are black to grayish, and the middle segments are black with pale green "arrow-shaped" spots pointed toward the tip. Segment 10 is black. The undersides and bottom of the sides of the abdomen are pale yellow or whitish. The face is olive with a black line. The eyes are bluish-gray. The femur is black with a pale streak along the bottom, and the tibiae are black. Males and females are colored similarly, but females have less extensive brown coloration than the males.

Olive Clubtails prefer mud-bottomed rivers and larger creeks. Unlike our other clubtails, Olive Clubtails often perch in trees, willows, bushes, and grass and hang up when at rest. They can be found further from shore than some of the other clubtails. Because of their pale green coloring and tendency to hang in vegetation, they can be easy to overlook.

The Olive Clubtail's range extends through the arid west from British Columbia south into central California. In Oregon, its known range is along the Columbia River, from its mouth to Hood River County, with only one record from Malheur County. One was photographed at Freeway Lakes in Albany, Linn County, in 2008.

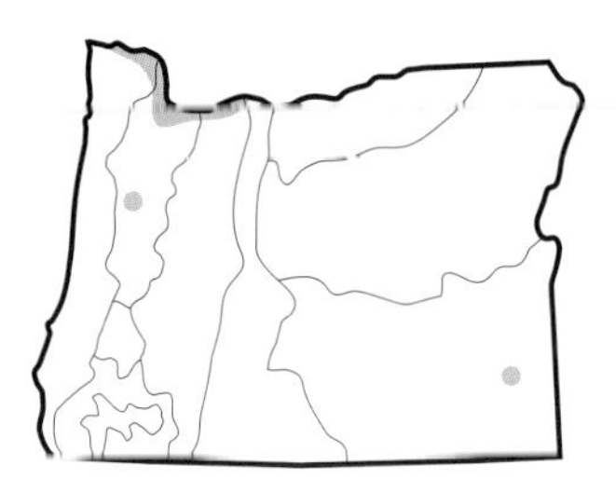

This is a relatively late-flying Gomphid with a flight season extending from early August into late September.

JAN	FEB	MAR	APR	MAY	JUN	JUL	AUG	SEP	OCT	NOV	DEC

Olive Clubtail male

Olive Clubtail female • photo by Ron Oriti

Long: 55–60 mm

Emeralds, Family Corduliidae

The emeralds are dark-bodied dragonflies named for their eye color. They have beautiful, emerald green eyes, which are jewel-like and come in contact with each other atop the head. The body has a metallic iridescence and the thorax can be quite hairy, often with spots. These coppery hairs may give the body a golden glow when seen in the right light. The anal loop in the hindwing is elongated but lacks a toe-like extension. The males are narrow-waisted at the base of the abdomen. The abdomen is flared, then tapers toward the tip, giving a spindle-shaped appearance. The emeralds are usually found at mountain lakes and boggy streams. They hawk along the shore and hover frequently. When they perch, they hang vertically or at an angle on stems or branches. The larvae are squat and hairy, and they sprawl in mud and detritus (Cannings, 2002).

Emerald larva (Baskettail) • photo by Giff Beaton

When patrolling, emeralds tend to dart along for a few meters and then hover, dart along for a few meters and then hover—then they will fly some distance and begin to repeat the darting and hovering behavior. When they hover in the sunlight seemingly looking at you, identification as an emerald will be obvious.

The emerald family is represented by three genera in Oregon: 1) the common emeralds, *Cordulia*, (one species, American Emerald, *C. shurtleffii*, the sole representative of this genus in North America); 2) the baskettails, *Epitheca*, (two Oregon species out of ten in North America); and 3) the striped emeralds, *Somatochlora*, (four Oregon species out of 26 in North America). The common emeralds have a forked lower appendage and no markings on the side of the thorax while the striped emeralds have pale or yellow markings on the side of the thorax. Unlike the other two genera, the baskettails have more subdued colors with blacks and browns marked with dull olive and butterscotch colors. The baskettails show small brown marks at the bases of the wings.

Identification Chart J compares five emerald species: The American Emerald in the common emerald genus, *Cordulia*, and four species in the striped emerald genus, *Somatochlora*. Identification Chart K compares the two similar species in the baskettail genus, *Epitheca*.

Identification Chart J • Emeralds: Common, Genus *Cordulia*, and Striped, Genus *Somatochlora*

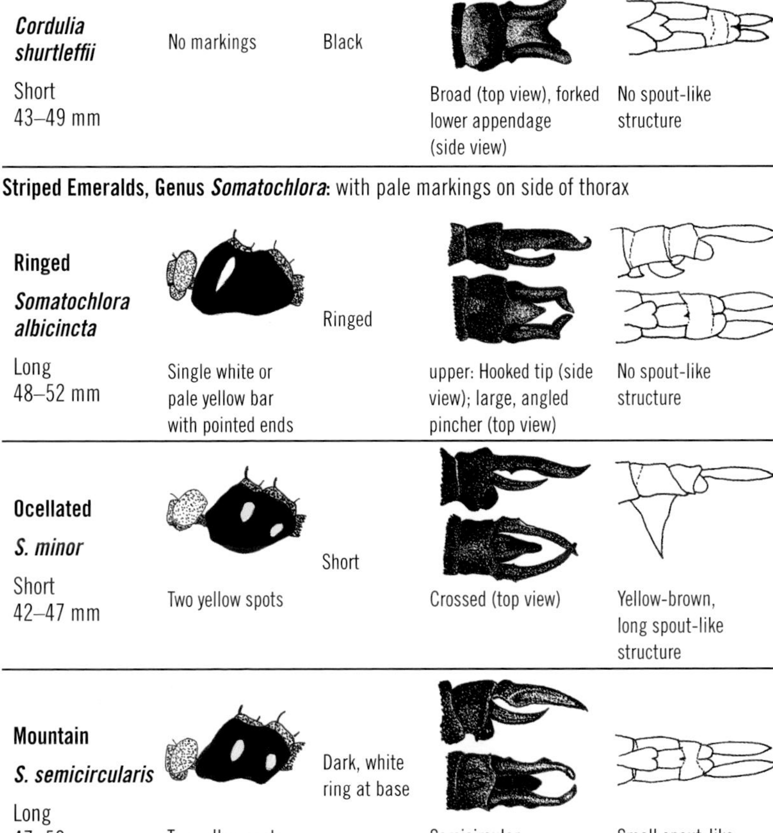

Species (size)	Thorax (side view)	Abdomen (notes)	♂ Appendages (side and top views)	♀ Subgenital Plate (side and bottom views)
American *Cordulia shurtleffii* Short 43–49 mm	No markings	Black	Broad (top view), forked lower appendage (side view)	No spout-like structure
Striped Emeralds, Genus *Somatochlora*: with pale markings on side of thorax				
Ringed *Somatochlora albicincta* Long 48–52 mm	Single white or pale yellow bar with pointed ends	Ringed	upper: Hooked tip (side view); large, angled pincher (top view)	No spout-like structure
Ocellated *S. minor* Short 42–47 mm	Two yellow spots	Short	Crossed (top view)	Yellow-brown, long spout-like structure
Mountain *S. semicircularis* Long 47–52 mm	Two yellow spots	Dark, white ring at base	Semicircular (top view)	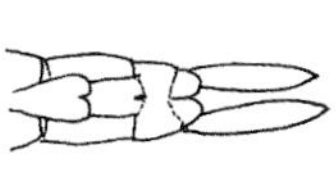 Small spout-like structure
Brush-tipped *S. walshii* Short 41–47 mm	 Stripe (front), oval spot (rear)	Small yellow side spots, S5–8	 Brushy, coppery hairs	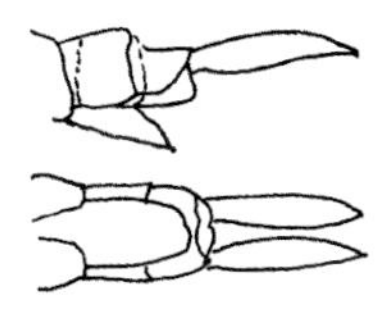 Dark, scoop-shaped, spout-like structure

American Emerald
Cordulia shurtleffii

The American Emerald has a dark face, and its hairy thorax is brown to metallic green and bronze with no pale spots or lines on the sides, which distinguishes this member of the common emerald (*Cordulia*) genus from the striped emeralds in the *Somatochlora* genus. It has a thin, white ring around the base of abdominal segment 3.

Its terminal appendages are quite short and the lower appendage has two lateral forks, a distinguishing feature. Females and males are colored and patterned similarly, but the female has a cylindrically-shaped abdomen.

The American Emerald is widely distributed in North America and its range extends southward in the mountainous west into California. In Oregon it is common up to 7,400 feet in elevation in forested mountain regions, except in the Great Basin portion of southeastern Oregon. It has been found at lower elevations, including coastal wetlands. It is the most common emerald that will be encountered in Oregon.

American Emeralds prefer mountain lakes, ponds, and bogs in forested regions and can be found in meadows and forest clearings. At rest, they hang vertically in trees and brush.

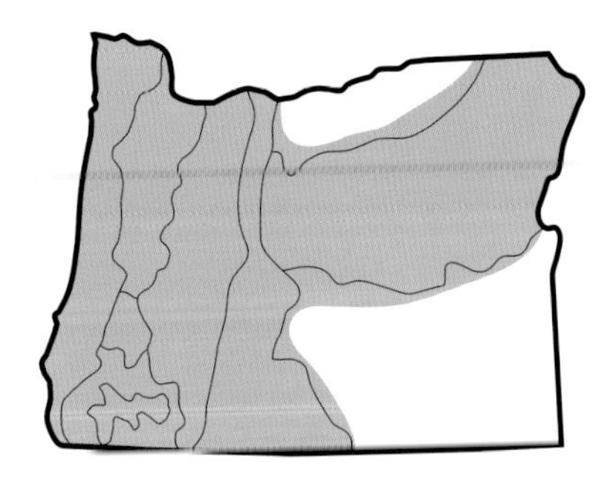

The adult flight season extends from early April to early September.

JAN	FEB	MAR	APR	MAY	JUN	JUL	AUG	SEP	OCT	NOV	DEC

American Emerald male

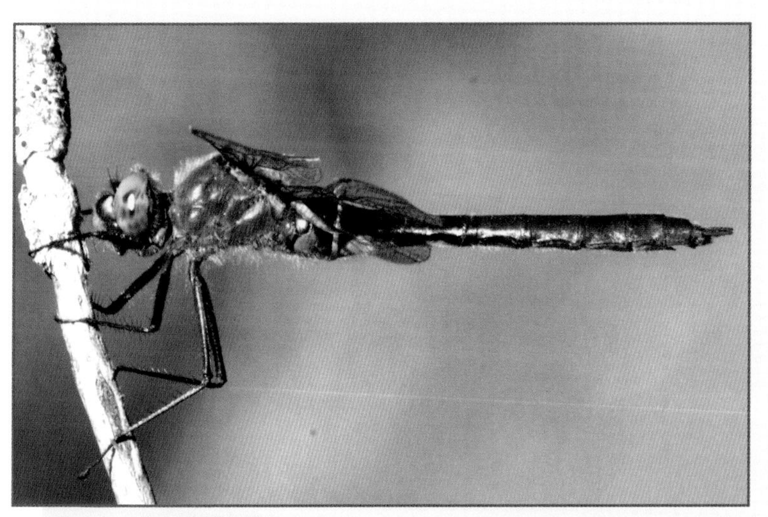

American Emerald female

Ringed Emerald

Somatochlora albicincta

Along with the Mountain Emerald, this is one of Oregon's two larger striped emeralds. The thorax is metallic green with one pale, yellow, thoracic side stripe with pointed ends. The abdomen is brownish-black with diagnostic, incomplete, narrow, white rings at the base of each segment. This identifying feature gives this species its names: the scientific name, *albicincta,* means girdled with white, and the English name, "Ringed." With experience, one can see the ringed abdomen even in flight, making identification in flight possible.

The male's upper appendages in top view extend parallel to each other then bend inward toward each other near the tip. In side view, the upper appendage has two teeth on the underside and the tip has a distinctive hook. The female structure is not obvious. In the bottom view, the subgenital plate is notched.

The Ringed Emerald is a transcontinental northern species with its range extending south from Canada into the Pacific Northwest as far south as Northern California. It is fairly common at high mountain lakes in Oregon's Cascade, Blue, and Siskiyou Mountains above 3,500 feet in elevation. It prefers open-water lakes with sparse emergent vegetation, but can be found foraging at adjacent small ponds, streams, and meadows, or hanging in nearby trees. The male tends to patrol low at the water's edge.

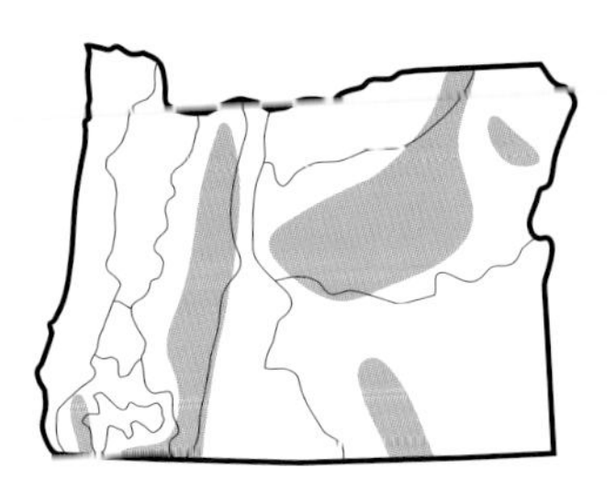

The adult flight season extends from late June to mid-September.

JAN FEB MAR APR MAY JUN JUL AUG SEP OCT NOV DEC

Ringed Emerald male

Ringed Emerald female

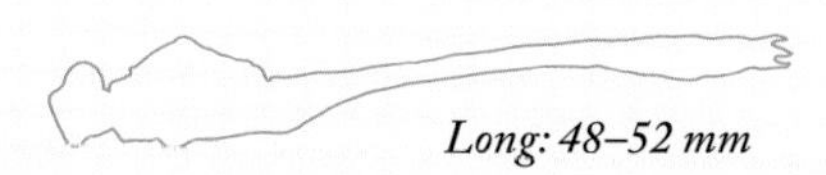

Long: 48–52 mm

Ocellated Emerald

Somatochlora minor

Along with the Brush-tipped Emerald, this is one of our two smaller striped emeralds. In fact, the Ocellated Emerald's scientific name, *minor*, refers to its small size. The dark green thorax shows two, oval, eye-like, yellow spots on its sides (thus the origin of its English name "Ocellated"). The overall appearance of the rather short abdomen is dark. There are smaller yellow spots atop the abdomen and yellow markings on the sides of the basal abdominal segments. The female and male are similar in color and pattern.

In side view, the male's upper appendages tip upward at the end like ski tips. In top view the upper appendages look like gently angled pincers that are pointed, and touch or cross at the tip. The female is unmistakable with an obvious, long, triangular, perpendicular subgenital plate.

The Ocellated Emerald has a transcontinental distribution, and its range extends south from Canada into the mountain west; it is not known from California. Its Oregon range is restricted to a few high-elevation streams in the central Cascades; at Crescent Creek in Klamath County at 4,620 feet in elevation, the most reliable Oregon location for this species, and near Todd, Irish, and Taylor Lakes. At Crescent Creek, males can be reasonably common patrolling the edges of the stream, but the females are very difficult to locate.

The Ocellated Emerald prefers clear streams flowing through wet meadows. They sometimes forage along forest edges and can be found hanging in trees. Females oviposit by tapping the water surface or laying eggs in mossy areas adjacent to the stream (Jones, 2008).

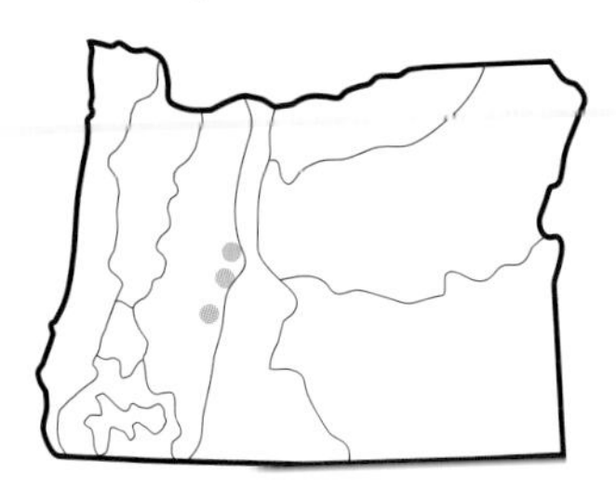

The adult flight period extends from late June to mid-August.

JAN	FEB	MAR	APR	MAY	JUN	JUL	AUG	SEP	OCT	NOV	DEC

Ocellated Emerald male

Ocellated Emerald female • photo by Ryan Brady (pbase.com/rbrady)

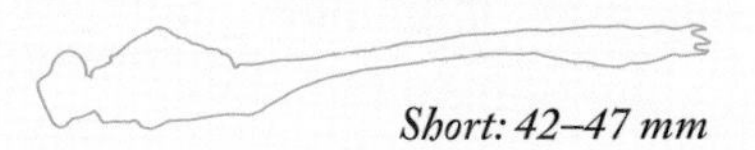

Short: 42–47 mm

Mountain Emerald

Somatochlora semicircularis

Like the Ringed Emerald, this is one of our two larger striped emeralds. The thorax is metallic green with two large, pale, yellow spots. While the Mountain Emerald appears similar to the Ringed Emerald, the two thoracic spots set it apart. Ringed Emeralds may show small, dull yellow spots on the sides of the mid-abdominal segments (segments 4–8). The abdomen does not have the white rings present in the Ringed Emerald. The female and male are marked similarly, but the female shows an orangish spot on the side of segment 3 near the top of the abdomen.

In top view, the male's upper appendages are curved in a semi-circular fashion (thus the species name "*semicircularis*") with the tips pointing inward toward each other, giving the two upper appendages a circular aspect. The subgenital plate is shallowly notched.

The Mountain Emerald is distributed in mountains in western North America and is common in Oregon's mountainous regions at higher elevations (above 2,800 feet). Mountain Emerald has been found at Gearhart Bog in Clatsop County near sea level.

Mountain Emeralds can be found at the edges of ponds, lakes, and boggy meadows where dense sedges are present. Males patrol these sedge areas, including seasonally dry areas, and may feed high and low in sunny clearings (Dunkel, 2000).

The adult flight season extends from mid-June to late September.

JAN	FEB	MAR	APR	MAY	JUN	JUL	AUG	SEP	OCT	NOV	DEC

Mountain Emerald male

Mountain Emerald female

Long: 47–52 mm

Brush-tipped Emerald

Somatochlora walshii

Along with the Ocellated Emerald, the Brush-tipped Emerald is one of Oregon's two smaller striped emeralds. The side of the thorax shows two pale marks, an elongated spot in the front and an oval spot in the rear. Faint yellow spots occur on the sides of the abdomen's basal segments. The female and male are colored and marked similarly.

In top view, the male's upper appendages are bent toward each other and are thick at the tip. In side view, the upper appendage shows two teeth on the underside and the tip is hooked upward in a ski-shape. The easily identifiable feature on the male is the thick tufts of long, coppery-brown hairs on the upper appendage that give this species its English name, "Brush-tipped." This "brushy" feature can be seen without the aid of a hand lens. The female's long subgenital plate is obvious and projects downward at a 45-degree angle. In bottom view, the female subgenital plate is slightly notched with rounded lobes.

The Brush-tipped Emerald's range is transcontinental and extends south from Canada into the northwest mountains as far south as Oregon. In Oregon, this species is not common, and has been found at higher elevation sites (from 3,200 to 4,900 feet) in the central and northern Cascades in Clackamas, Linn, Lane, and Klamath Counties. We have found it at Gold Lake Bog in Lane County, and Gordon Lakes in Linn County.

The Brush-tipped Emerald can be found in dense sedge wetlands with small, clear streams running through them. Look for them near lake outlets. The female lays eggs by tapping her abdomen below the water surface (Jones, 2008).

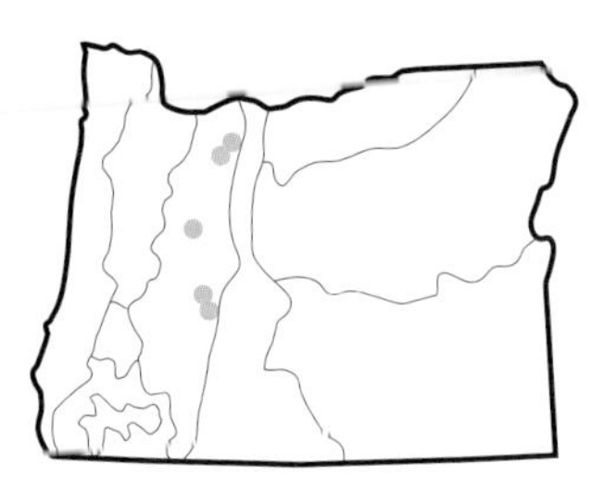

The adult flight season extends from early July to late September.

JAN FEB MAR APR MAY JUN JUL AUG SEP OCT NOV DEC

Brush-tipped Emerald male

Brush-tipped Emerald female (exuding egg batch)

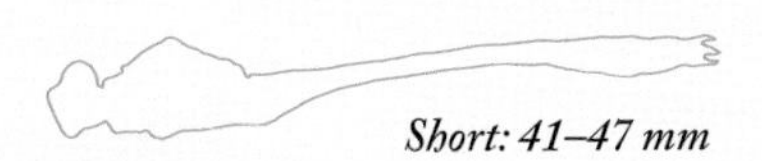

Short: 41–47 mm

Mountain Emerald male

Beaverpond Baskettail male

Baskettails

Baskettails (genus *Epitheca*) are subtly brown colored with muted yellow markings and a very hairy thorax. They frequent riparian zones, still backwaters and ponds, including beaver ponds, and along forest edges and openings, including trail and road cuts. Both of our species have early season flight periods and can occasionally be found in mixed swarms. Our two species appear almost identical in the field, and can be told apart by comparing the terminal appendages and top of the head for presence or absence of a T-Spot (see the Baskettail Identification Chart below).

Identification Chart K • Baskettails, Genus *Epitheca*

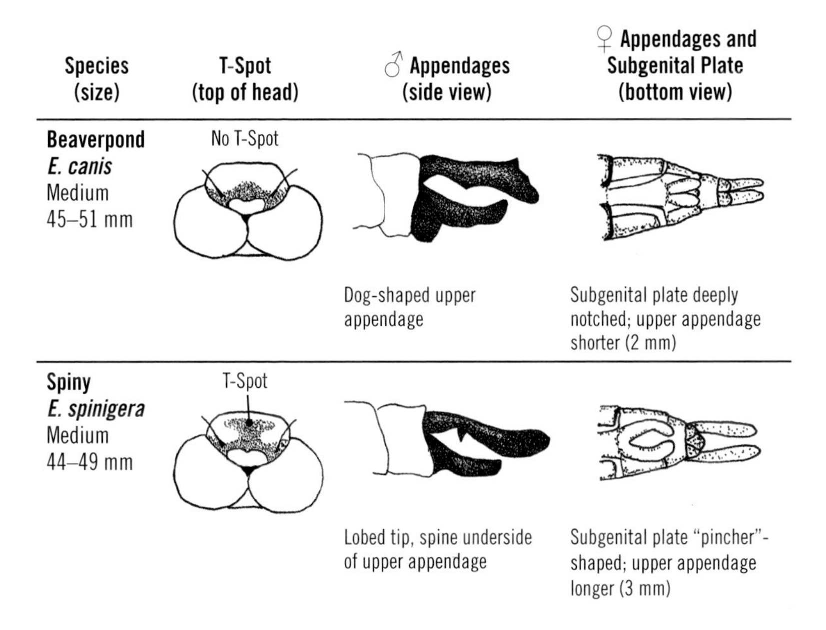

Species (size)	T-Spot (top of head)	♂ Appendages (side view)	♀ Appendages and Subgenital Plate (bottom view)
Beaverpond *E. canis* Medium 45–51 mm	No T-Spot	Dog-shaped upper appendage	Subgenital plate deeply notched; upper appendage shorter (2 mm)
Spiny *E. spinigera* Medium 44–49 mm	T-Spot	Lobed tip, spine underside of upper appendage	Subgenital plate "pincher"-shaped; upper appendage longer (3 mm)

Beaverpond Baskettail

Epitheca canis

Beaverpond Baskettails can be found at beaver ponds, as well as other suitable habitats. The thorax of this species is brown and very hairy. The abdomen is flat, narrow at the base and tip, and broad in the midsection—with sides that show a series of golden-yellow spots which vary with maturity. The eyes are bluish-green at maturity.

Unlike the Spiny Baskettail, the Beaverpond Baskettail has no T-spot atop its forehead. The wings are clear except for small dark patches at the base. The male's upper appendage, when viewed from the side, has a "dog head" shape on the tip; thus its species name, *canis.* The female's subgenital plates are deeply notched and the abdominal appendages are short (about 2 mm long) relative to the Spiny Baskettail female appendages, which are about 3 mm long.

The Beaverpond Baskettail is transcontinental and its western range extends from Alberta into the northern half of California. In Oregon, its range is primarily western, up to 4,500 feet in elevation, with records from the eastside slopes of the Cascades in Deschutes and Klamath Counties. It is more often associated with streams than is the Spiny Baskettail.

This species perches vertically, frequently in low vegetation, and can be approached easily— we have captured individuals with our bare hands. Males are most often encountered patrolling over water. They usually fly along a relatively short section of shoreline, then out toward the center of the pond and back to the starting point where they begin again. The female lays her egg mass in a sticky string or ball and taps the string or mass onto the surface to release them into the water at ponds and slow-moving streams (Dunkle, 2000).

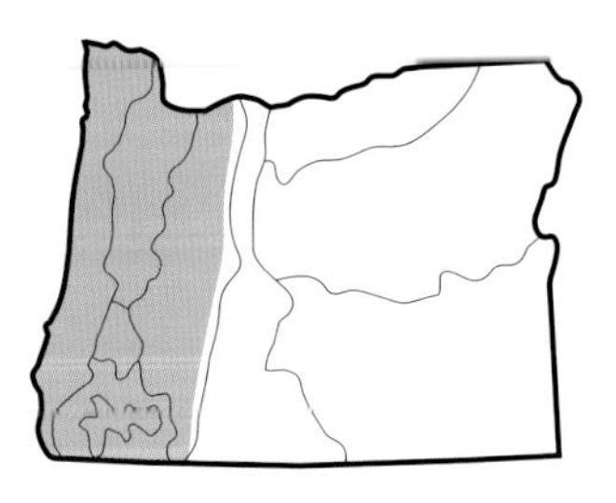

The adult flight season is relatively early and short, extending from early May through late July.

JAN	FEB	MAR	APR	MAY	JUN	JUL	AUG	SEP	OCT	NOV	DEC
				▬	▬	▬					

Beaverpond Baskettail male

Beaverpond Baskettail female

Medium: 45–51 mm

Spiny Baskettail

Epitheca spinigera

The Spiny Baskettail is brown overall with a very hairy thorax and light yellow markings on the side of the abdomen. Its body color, eye color, and shape are similar to the Beaverpond Baskettail. Baskettails are difficult to identify to species without close inspection of the T-spot and appendages.

In the hand, the Spiny Baskettail has a dark T-spot atop its forehead. The male's upper appendage, when viewed from the side, is a simple, long lobe with a spike projecting downward about a third of the way along the upper appendage's length. However, this spike may be reduced or lacking in some individuals. The female Spiny Baskettail's subgenital plate is pincher-shaped and her appendages are longer (about 3 mm long) relative to the female Beaverpond Baskettail appendages (about 2 mm long).

The Spiny Baskettail's range is transcontinental, and in the west, its range extends south from Alberta into northeastern California. In Oregon, its range is predominately western with records from the eastern side of the Cascades in Deschutes and Klamath Counties. It is more often found at ponds than at streams.

The Spiny Baskettail hunts over trails and cleared areas, but often at tree top level with frequent forays over adjacent ash and willows. Males often seem to patrol along a treeline rather than over water. This species seems to stay just above net's reach much of the time.

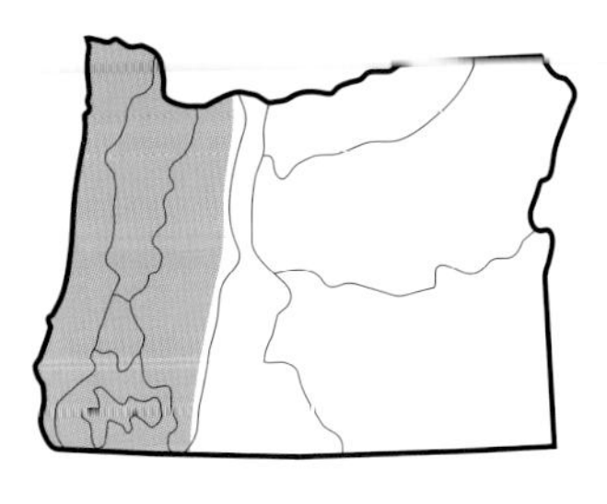

The adult flight season is relatively early and extends from mid-April to early July.

JAN	FEB	MAR	APR	MAY	JUN	JUL	AUG	SEP	OCT	NOV	DEC

Spiny Baskettail male

Spiny Baskettail female

Medium: 44–49 mm

Hoary Skimmer female

Western Pondhawk eating leafhopper

Skimmers, Family Libellulidae

The family Libellulidae is large and diverse with 109 North American species. The Libellulids include many of Oregon's most common and widespread species. The genus name, *Libellula*, means little book, referring to wings folded like pages of a book. The 31 Oregon species in this family fall into ten different genera and are some of the most highly patterned and easily recognized dragonflies. Many are found throughout the state and have a long flight season making them some of the species most commonly observed.

Oregon has a single species in both the *Erythemis* and *Pachydiplax* genera. Both species are common and blue males are similar enough to be confused on first sight. *Pachydiplax* is a monotypic genus; that is there is only a single species in the genus. Our only species of *Ladona*, the corporal genus, has a restricted montane range. However it can be quite common and unmistakable at some Cascade Lakes where it favors perching on logs and the ground. There are four species of small, dark-colored whiteface (genus *Leucorrhinia*, see Identification Chart L) found in Oregon, with three of these being montane species. However, one species, the Dot-tailed Whiteface, is common in lowland wetlands. Among the eight species of *Libellula*, the king skimmers are some of our most common species even in urban areas. Many species have patterned wings and are greatly appreciated by people visiting our wetlands and parks (see Identification Chart M). The whitetails (genus *Plathemis*) are represented by two species. While the Common Whitetail is found throughout the state, the Desert Whitetail is found only in Oregon's southeastern high desert. These medium-sized dragonflies with bright pruinose white abdomens are the only

Comanche Skimmer male

two species in this genus. The *Paltothemis* genus (Red Rock Skimmer) has only been found a few times at a single stream in southeastern Oregon. The males are bright red-and-black patterned dragonflies, while the females are brown and black. Rainpool gliders, the genus *Pantala*, are represented by two occasional visitors to the state. Both have very long, broad wings, aiding these dragonflies in their seasonal migrations. The *Tramea* genus is represented by the Black Saddlebags with its unmistakable black spot on the hind wing. This large, dark-colored dragonfly is common even in urban areas. The meadowhawks (*Sympetrum*) are small, red-patterned dragonflies and are common throughout the state (see Identification Chart O). They are a diverse group represented by ten species in Oregon, and they can be difficult for beginners to identify. Indeed, dull-colored females can be difficult even for the experienced!

Most of the species in this family are perchers, spending much of the time perched on vegetation, rocks, and on the ground. Some, such as males of the Cardinal Meadowhawk (*Sympetrum illotum*), Western Pondhawk (*Erythemis collocata*), and Blue Dasher (*Pachydiplax longipennis*) species, prefer perches located over the water. Perchers will obelisk during hot weather as a method of regulating body temperature. In order to minimize the surface area exposed to the sun, they perch with the abdomen pointed at the sun. This produces an amazing acrobatic posture that is surprising to see. Perchers in this family fly out to defend a territory or feed, often returning to the exact same perch. You may see some species, such as the Cardinal Meadowhawk (*Sympetrum illotum*), using the same perch day after day. In fact, a good perch site will often be occupied by different individuals of the same species from year to year.

The gliders (*Pantala*) and saddlebags (*Tramea*) are usually seen flying, much of the time overhead, but also patrolling a few feet above ponds. The gliders are strong fliers, have large, broad hind wings, and migrate great distances, often with weather fronts. They are found worldwide, except Antarctica. In North America they are known to migrate northward from tropical areas to breed, and to migrate southward in the post breeding season, and we have seen them flying far from water in the deserts of Mexico. Saddlebags often wander away from water and are seen in our yards and gardens. Often they are noticed because the moving flight shadow on the ground causes an instinctive look upwards.

Pruinescence is an attribute exhibited by some male Libellulids when mature. This appears as a powdery coating—often on the abdomen, but may also be on other areas of the anatomy—and can be various colors depending on the species. This pruinescence has been found to reflect ultraviolet light, which is not detected by the human eye but is visible to many other animals (Corbet, 1999) and likely functions in territorial and sexual display.

After mating, most females oviposit by flying just above the water and continually dipping the abdomen to release eggs. The eggs released can be clusters containing large numbers of eggs. Egg laying has been measured in numbers exceeding 1,000 eggs per minute in some Libellulids (Corbet, 1999). This is impressive when considering that the eggs are fertilized as they pass through the oviduct. In some cases the female is guarded by the male while oviposition takes place. The meadowhawks often engage in contact guarding remaining attached during oviposition. The Striped Meadowhawk (*Sympetrum pallipes*) frequently oviposits in groups in dry areas that will later flood, providing appropriate habitat after winter rains. In some of the skimmers, the male simply flies in circles above the ovipositing female, driving off other Odonates. The Black Saddlebags (*Tramea lacerata*) has a distinctive ovipositing behavior. The male and female remain attached and fly above the pond a few feet. Occasionally, the male releases the female and she glides down to the surface to dip the abdomen and drop eggs. She then flies back up and the male reattaches. This behavior is repeated over and over again. One wonders at the coordination of this effort, and the signals that must be passed to trigger it.

Migrations or directed flights of the Variegated Meadowhawk *(Sympetrum corruptum)*, are sometimes noted on the Oregon coast. There is not much known about this behavior but it seems to be initiated by cold fronts causing a southward movement. In one case, thousands of individuals were filmed at the same spot moving south for five days near Yachats.

Skimmer family larva

The larvae of the Libellulids are sprawlers and climbers, lying in wait for prey on the bottom or in vegetation in ponds, bogs, lakes and sluggish streams.

We have altered the sequence of the traditional presentation so that several groups with similar characteristics or appearance can be compared more easily. For example, the Western Pondhawk and Blue Dasher are presented together because they are both smaller blue skimmers. The Flame Skimmer and Black Saddlebags are so unique that they are not included in any of the Identification Charts or Figures in this section.

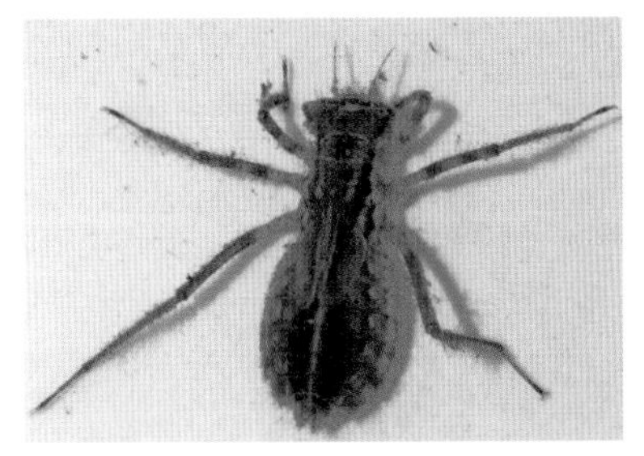

Skimmer larva

Bleached Skimmer male

Western Pondhawk

Erythemis collocata

The Western Pondhawk has clear wings and a powdery blue, pruinose thorax and abdomen. The thoracic sides are unmarked, pruinose blue. The eyes are dark blue and the face is green. Western Pondhawk females and the immature male are a beautiful and unique lime green color and easy to identify by sight; they show a dark line down the top of the abdomen. The immature male gradually changes to the blue pruinose color, so it is not uncommon to see males with some light green on the upper thorax. The female has a thicker abdomen than the male.

The male Blue Dasher is similar, but has a white face, marked thoracic sides, and an orange tint at the wing bases. The female is dark and marked in yellow, not green, with dark markings like the female Western Pondhawk.

Worldwide there are ten species in this genus, with seven in North America and just one species in Oregon. The Western Pondhawk is found in the western United States and ranges from southern British Columbia into Mexico. In western and northern Oregon it is common in unforested areas below 1,000 feet in elevation and up to 4,900 feet in the Basin and Range region. It is scarce in forested mountain regions.

The Western Pondhawk is one of the few skimmers to perch directly on the ground. It can also be found perching horizontally on low or floating vegetation at many kinds of ponds, lakes, and marshy areas.

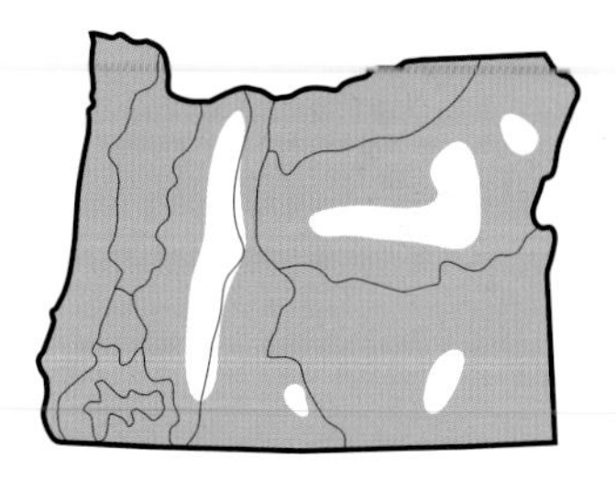

The adult flight season extends from late April to early October.

JAN	FEB	MAR	APR	MAY	JUN	JUL	AUG	SEP	OCT	NOV	DEC

Western Pondhawk male

Western Pondhawk female

Short: 39–42 mm

Blue Dasher

Pachydiplax longipennis

Only one species of the genus *Pachydiplax* exists in Oregon, North America, and the World. The Blue Dasher has a thorax and abdomen of predominately pruinose blue, but the sides of the thorax retain black with greenish stripes. The blue abdomen has a dark tip and abdominal appendages. The wing bases show amber patches with short dark streaks. The eyes are metallic green, and the face is white with a black spot on the top. Look for the white face and striped thorax to distinguish perched males from the Western Pondhawk. The female and juvenile male have reddish-brown eyes on top, black and orangish-yellow stripes on the side of the dark thorax, and paired rows of yellow dashes on the top of the dark brown abdomen. The Latin species name, *longipennis,* suggests long wings, but the wings are not remarkably long in proportion to body length (Needham, et. al., 2000).

The Blue Dasher has a transcontinental distribution and ranges from southern British Columbia to Belize. Several sources indicate that this species is expanding its range northward. It is common in western Oregon below 1,000 feet in elevation and in the Columbia Basin. In the remainder of eastern Oregon, it is scarce.

The Blue Dasher can be found at ponds, small lakes, and slow moving creeks with emergent or floating vegetation. It usually perches with wings held forward and downward on plant stems and twigs, often over water. Unlike the Western Pondhawk, the Blue Dasher does not perch on the ground.

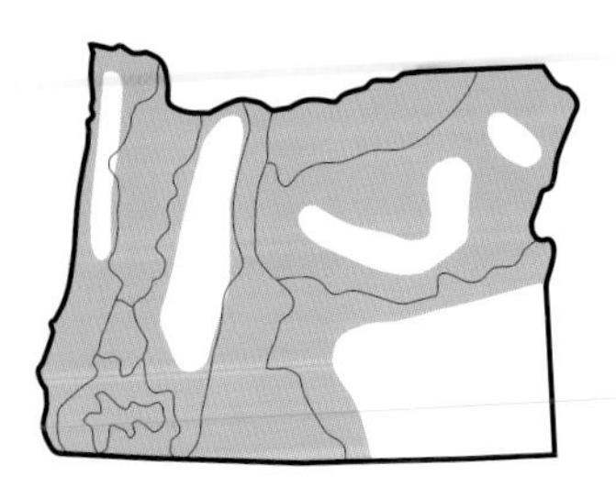

The adult flight season lasts from mid-May to early October.

JAN	FEB	MAR	APR	MAY	JUN	JUL	AUG	SEP	OCT	NOV	DEC

Blue Dasher male

Blue Dasher female

Very Short: 35–42 mm

Chalk-fronted Corporal

Ladona julia

The Chalk-fronted Corporal's English name is based on the two light-colored, parallel bars on the front of the thorax, which resemble the insignia for the military rank of Corporal. The stocky body is dark brown to black, with those light parallel bars and base of the abdomen becoming pruinose blue on the mature male. The tip of the abdomen remains dark. The wings are clear except for brown patches at the base of each wing and a dark pterostigma. The eyes are dark brown and the face is black. The female is brown: tan face and brown eyes, and a warm brown body with a dark abdominal tip. Some older females gain pruinosity. The male and female wing pattern is similar.

Chalk-fronted Corporals can be found at lakes, ponds, bogs, and slow streams with marshy shores in forested areas. They often perch on floating logs, sticks, rocks, or the ground. Away from water, they often perch low in sunny aspects on tree trunks and fallen logs. They hover briefly and then dart rapidly low over the water. Males hover-guard females during oviposition.

The Chalk-fronted Corporal is transcontinental and its range extends from Alberta and British Columbia south into central California. In Oregon it is found in mountainous habitats in the Cascades and Klamath Mountains from 2,500–5,000 feet in elevation—with no records from the Blue Mountains. We have found it to be most common at Otter Lake in the Three Sisters Wilderness of Lane County, where it is almost as numerous as the mosquitoes!

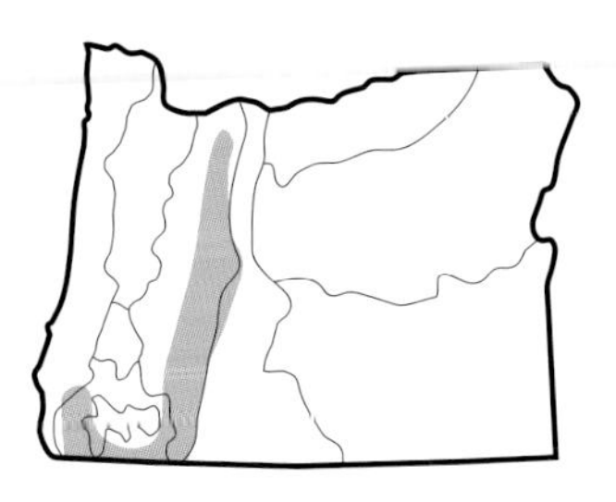

The adult flight season extends from late May to early September.

JAN	FEB	MAR	APR	MAY	JUN	JUL	AUG	SEP	OCT	NOV	DEC

Chalk-fronted Corporal male

Chalk-fronted Corporal female

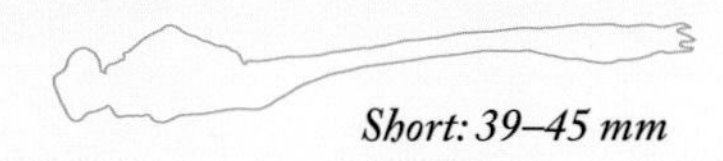

Whitefaces, Genus *Leucorrhinia*

The whiteface species are attractive, relatively short (33–39 mm), dark-bodied dragonflies showing prominent white faces; the genus name means "white nose." The dark bodies are marked in beautiful patterns of red or yellow on the thorax and abdomen. They have black legs and a dark patch at the base of the hind wing. The pterostigma is dark, short, and relatively wide with two cross-veins below it.

Four of the seven North American whiteface species have been recorded in Oregon. They are not to be confused with other Oregon dragonflies with white faces: White-faced Meadowhawk, which has a red body, or three blue-bodied species: Blue Dasher, and two larger, Great Basin region skimmers—the Bleached and Comanche. Identification Chart L illustrates and summarizes the key features for identifying the four whiteface species.

Whitefaces are gorgeous when viewed at close range, and they are generally easy to approach and observe. In the hand, their colored patterns are even more evident. However, by remaining low in emergent plants or on wet ground, these strikingly marked dragonflies blend perfectly into the shadowy background.

The whiteface genus takes some study to identify at the species level. Juvenile males begin with yellow body markings on the thorax and abdomen that turn red in maturity. Most females retain yellow body markings, but some may turn red at maturity. The details of the markings may vary and are not the most reliable diagnostic characters for identification to the species level.

The male Dot-tailed Whiteface is easy to identify with the unaided eye. It is also the only species found in lowland elevations and often uses higher perches like cattails or reed canary-grass. The other three species of boreal males are red and black and are found in higher elevation situations. Male Hudsonian Whitefaces can often be identified by the red marks atop the abdomen, but the abdominal spots are lacking in some individuals. For any individual lacking abdominal spots, we recommend examination in the hand for positive identification. The juveniles and females are difficult to separate without careful study in the hand, and the male Crimson-ringed and Belted Whiteface are nearly identical.

The Wonder of Dragonflies

Hudsonian Whiteface male with worn wings

Identification Chart L • Whitefaces, Genus *Leucorrhinia*

Species (size)	♂ Abdomen (top view)	♂ Hamule (side view)	♂ Appendages (side and bottom views)	Wing (radial planate)	♀ Vulvar Lamina (bottom view)
Black abdomen with distinct yellow spot atop Segment 7:					
Dot-tailed ***Leucorrhinia intacta*** Short 34–35 mm	Black with pair of large yellow spots atop S7	 Wide; hooked at tip; inverted "J"	lower: Forked; tips widen outward	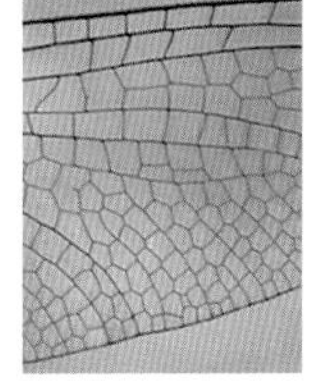 Two rows of cells	Two small, widely-spaced lobes
Black and red (or yellow) pattern on Thorax and Abdominal Segments 1-3:					
Hudsonian ***L. hudsonica*** Short 27–35 mm	Red spots atop S4–7	 Inverted "J"	lower: 3/4 length of cerci	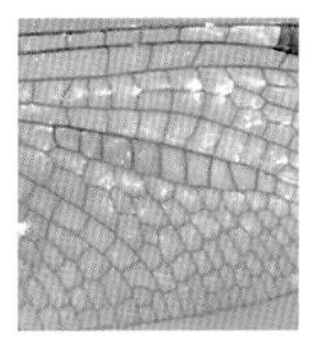 One row of cells	Large lobes
Crimson-ringed ***L. glacialis*** Medium 33–36 mm	Black with no spots	 Hooked tip	 lower: 1/2 length of cerci	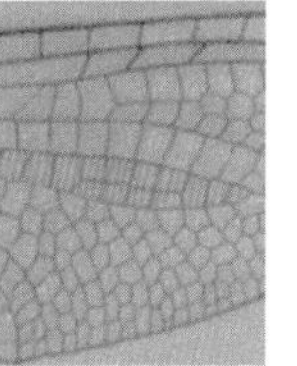 Typically, two rows of cells	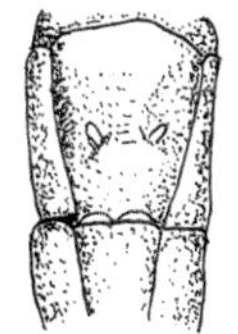 Short lobes separated at base
Belted ***L. proxima*** Long 34–39 mm	Black with no spots	 Curved and thick; "Crescent"-shaped	 lower: 2/3 length of cerci	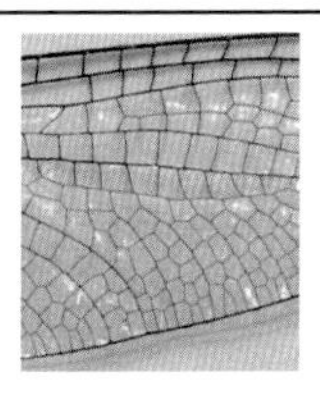 One row of cells	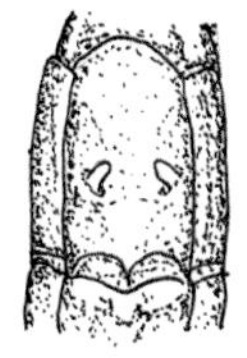 Short, rounded lobes touch at base

Crimson-ringed Whiteface

Leucorrhinia glacialis

This is a beautiful, small, black-and-red whiteface found at high elevations. Like the other whitefaces, it is primarily black with black legs and a white face. The mature male abdomen is black with red at the base and no red spots atop abdominal segments 4–10. The red on the abdomen extends forward between the wings. Where the mature male is red, the immature male is yellow. The female is patterned like the other female whitefaces (spots atop the abdomen), and her markings may be yellow or red.

The larger size and lack of red spots atop abdominal segments 4–10 make separation from Hudsonian generally easy, with the exception of those Hudsonian lacking the abdominal spots. Separating Crimson-ringed from Belted requires examination in the hand: the Crimson-ringed usually shows two rows of some cells in the radial planate, and the Belted shows one row of cells. The male's lower appendage (epiproct) is relatively short and is half the length of the upturned upper appendages (cerci); whereas the Belted's epiproct is proportionately longer. The male's hamule has a small hook at its tip, whereas the Belted's has no hook and has a crescent shape. The female Crimson-ringed's vulvar laminae show two very small lobes separated by a small gap. The lobes do not touch at their bases.

Like the other whitefaces, the Crimson-ringed perches on the ground or on low vegetation. Males defend very small territories of 1–2 square yards (Dunkel, 2000). The males guard the females as they oviposit in shallow water among emergent vegetation.

This species range extends across Canada and the northern tier states and south into Wyoming and Nevada. It can be common on the fringes of ponds and small lakes with emergent and floating vegetation above 4,000 feet elevation in the Cascade Range and some Eastern Oregon mountains.

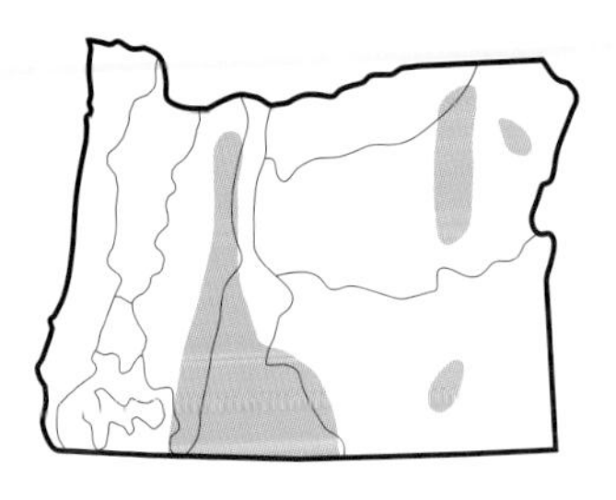

The adult flight period extends from early June to early September.

JAN FEB MAR APR MAY JUN JUL AUG SEP OCT NOV DEC

Crimson-ringed Whiteface male

Crimson-ringed Whiteface female

Medium: 33–36 mm

Hudsonian Whiteface

Leucorrhinia hudsonica

This black-and-red whiteface is smaller than the Crimson-ringed or Belted Whitefaces. Like the other members of this genus, it is black with a white face. The Hudsonian male has deep, rich, complex red markings on the thorax and red spots atop abdominal segments 1–7. The Crimson-ringed and Belted Whitefaces lack red spots atop segments 4–7. In the Hudsonian males, the dorsal red markings may be faded or, in rare cases, absent, making hand examination the only way to make an absolute identification. The females, immature males, and mature males are patterned similarly, but in females and some juvenile males, the spots and markings are yellow. Females may also be red like the mature male.

In the hand, check the top of the abdomen for any inconspicuous red spots. The wing has one row of cells in the radial planate (see Identification Chart L). The female's subgenital plate has the largest lobes of any of the four whiteface species. The female's dark basal hindwing markings may include pale veins, a unique feature (Dunkle, 2000). The costa is yellowish beyond the nodus, and in the male is bright yellow around the pterostigma (Needham, et. al., 2000). The male's lower appendage (epiproct) is widely notched. The rear arch on the male hamule has an inverted "J" shape.

Males defend very small territories (Dunkle, 2000) and guard females ovipositing among emergent vegetation in shallow water (Manolis, 2003).

The Hudsonian Whiteface is named for its habitation in the Hudsonian Life Zone, and its type locality is on Hudson Bay, New Brunswick. It is transcontinental in its distribution, and its range extends south into northern New Mexico. It can be found commonly along vegetated shores of ponds, bogs, and lakes, and in wet meadows above 3,000 feet in elevation in the Cascade Range and eastern Oregon mountains. It perches on the ground or in low vegetation and can be approached closely.

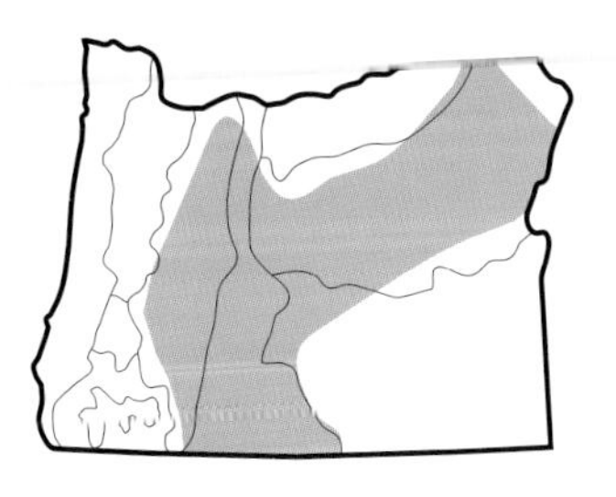

The adult flight season extends from mid-May to mid-September.

JAN FEB MAR APR MAY JUN JUL AUG SEP OCT NOV DEC

Hudsonian Whiteface male

Hudsonian Whiteface gynochrome female (inset: androchrome female)

Short: 27–35 mm

Dot-tailed Whiteface

Leucorrhinia intacta

This small dragonfly is black with a white face and a prominent square, yellow spot atop abdominal segment 7. This squarish spot, which is the source of its name, is most obvious on the mature male, which shows no other yellow markings atop the abdomen. Although the females and immature males show additional yellow spots atop the middle segments of the abdomen, the squarish segment 7 spot is still very obvious. Other features on both sexes are black legs, dark markings at the base of each wing, and bodies covered with hairs, many of which are white. The male also shows a large, yellow spot on the basal side of the abdomen. Females and juvenile males have additional yellow spots along the side of the abdomen, but in the males, the spots fade to black as they mature.

The male's lower appendage (epiproct) is forked and the tips spread outward. The female's vulvar laminae exhibit two widely spaced, parallel, short lobes that look like two tiny bumps. The rear edge of the male hamule is vertical.

The Dot-tailed Whiteface's range is transcontinental and ranges south from Canada to New Mexico. Unlike the other three species in this genus, it is widespread in Oregon and can be found in abundance at lower elevations from the coast, Coast Range, valleys, and foothills, and less commonly up to the highest mountain lakes.

The Dot-tailed Whiteface can be found at the edges of well-vegetated ponds (including created farm and garden ponds), lakes, and in slow moving river side-channels. The males perch in sunny exposures on the ground or low vegetation and may also be seen perching on emergent or floating vegetation (like water lily or pondweed leaves). Males guard females by hovering nearby and chasing away competitors (Manolis, 2003).

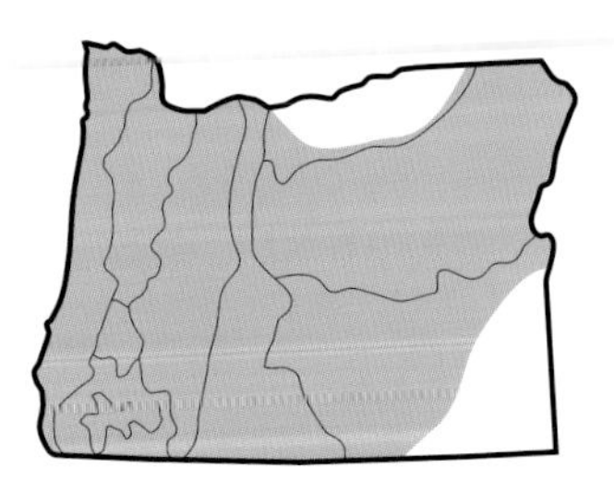

In Oregon, the adult flight period extends from early May to mid-August.

JAN	FEB	MAR	APR	MAY	JUN	JUL	AUG	SEP	OCT	NOV	DEC
				▬	▬	▬	▬				

Dot-tailed Whiteface male

Dot-tailed Whiteface female

Short: 34–35 mm

Belted Whiteface

Leucorrhinia proxima

The Belted Whiteface was formerly named the Red-waisted Whiteface. It is another small, black whiteface with the male showing red markings on the thorax and base of the abdomen, and no red markings atop abdominal segments 4–10. This species is the longest among the three Oregon black-and-red boreal whitefaces. Like the other whiteface species, the immature male is yellow where the mature male is red. The female may have yellow or red markings.

In the hand, the Belted Whiteface has one row of cells in the radial planate, whereas the Crimson-ringed usually has two rows (see Identification Chart L). The lower appendage of the male is parallel-sided and two-thirds the length of the upper appendages, whereas the Crimson-ringed is shorter proportionately (about half the length of the upper appendages) (Dunkle, 2000). The female's subgenital plate has very short, rounded lobes that are widely notched, touching at their bases. The hamule of the male Belted Whiteface has a rear arch that is curved in a "C" or crescent shape and has no hook at its end, whereas the Crimson-ringed's hamule has a hooked end.

The Belted Whiteface frequents the boggy and marshy edges of high-elevation ponds and lakes. The Belted Whiteface's range spans North America across Canada and the northern tier states into Alaska. In Washington it is common in the northern mountains (Paulson, 1999). In California it is known from one sphagnum bog containing floating islands at 5,500 foot elevation at Willow Lake in Plumas County (Manolis, 2003). In Oregon, it was first recorded by the authors in 2008 along the shady, marshy shores of Magone Lake at 4,907 feet elevation in the Blue Mountains, Grant County. It is common here and was found perching low in brush along the shore, on floating logs, and on the ground.

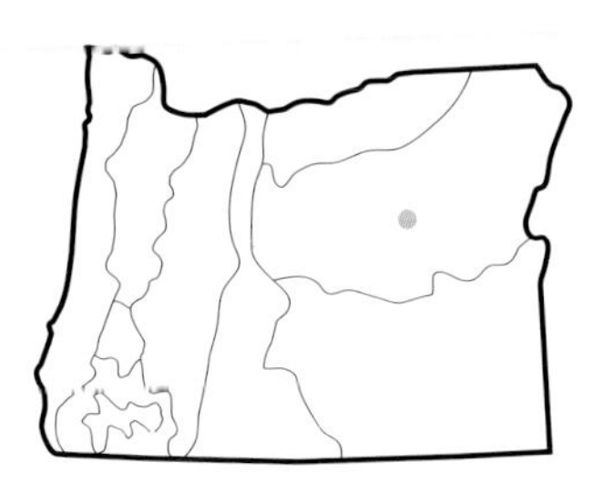

Adults have been recorded in Oregon only during July, but this will likely expand with more observations. In California, the flight season is from mid-June to mid-August (Manolis, 2003), and in Washington, the flight season is from mid-June to mid-September (Paulson, 1999).

JAN	FEB	MAR	APR	MAY	JUN	JUL	AUG	SEP	OCT	NOV	DEC
						▬					

Belted Whiteface male

Belted Whiteface female

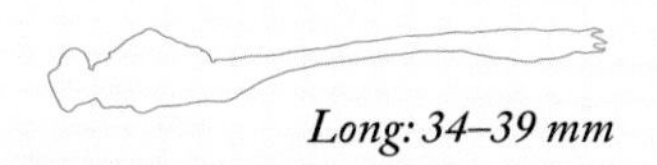

Long: 34–39 mm

Comanche Skimmer

Libellula comanche

At maturity, the large, male Comanche Skimmer attains a blue pruinosity giving its thorax and abdomen a pale, blue, unmarked color. It has a uniquely bi-colored pterostigma which is white on the inner three-quarters and dark on the outward margin. The wings are clear with a black costa. It has pearly blue-gray eyes and a creamy-white face. The legs are black. These features make the Comanche Skimmer easy to identify and separate from other blue dragonflies. The female also has blue-gray eyes and a white face, but the white portion of the pterostigma may be brown on older females. Its costa is also black, but the wing tips are often tinged with brown. The female's body color is overall brown with broad whitish patches on the sides of the thorax separated by a narrow dark stripe. The abdomen, brown and pale yellowish, has a wide, dark, median line. The abdomen has a double row of light, broken stripes on either side of this dark median line tapering on the rear segments. Immature males are colored like the females, but show a bi-colored pterostigma like the adult male.

Male Comanche Skimmers tend to perch at eye level in vegetation and actively defend territories by chasing other males, including males of other dragonfly species. In Oregon, the Comanche Skimmer has been found only at hot springs and a hot-spring-fed pond. In California it is also found in foothill springs and most of these springs and pools exhibit dense emergent vegetation, often including cattails (Manolis, 2003).

The Comanche Skimmer is a southwestern species ranging from easternmost Texas westward and north into southern Oregon and Idaho. Its range extends south into Mexico. In Oregon, it is known from only two locations in the Harney County Great Basin region: Borax Lake and Twin Springs. It is common at both locations. It is worth the trip to see this spectacular dragonfly.

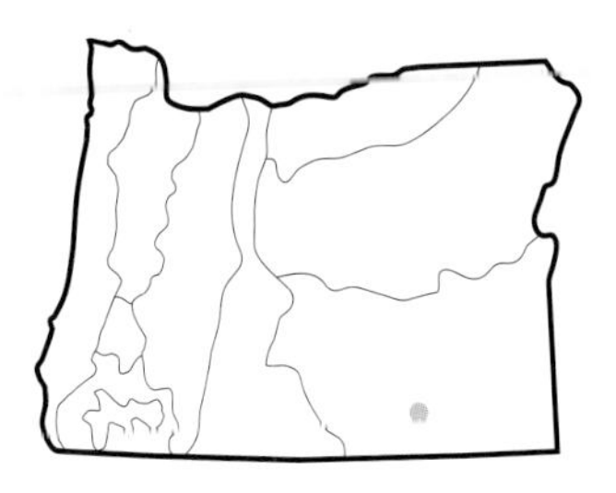

The adult flight season extends from late May to mid-September.

JAN	FEB	MAR	APR	MAY	JUN	JUL	AUG	SEP	OCT	NOV	DEC

Comanche Skimmer male

Comanche Skimmer female (inset: side view)

Long: 47–55 mm

Bleached Skimmer

Libellula composita

Tim Manolis (2003) described the Bleached Skimmer as a "ghostly apparition on the wing," an apt description of this pale, arid country dragonfly. Mature males at full pruinosity have a pale blue thorax with pale sides and a pale blue abdomen with dark tip. The blue pruinosity on the abdomen obscures underlying brown and yellow patterns. The face is creamy white and the eyes are lustrous whitish-blue. The wings have a pale white costa, a small brown spot at each nodus, a brown patch at each base, and a black pterostigma. We have also observed males lacking the brown nodal spot. The wing pattern is unique among Oregon dragonflies. The legs are pale at the base and otherwise black.

While some females attain pruinosity, most females are yellow and brown. The distinctive wing pattern is similar to the males, but the nodal dark spot may be absent. The female's thorax is light yellow with brown stripes on the front, and the yellow sides are broken with two dark stripes. The abdomen has a wide, dark medial line bordered by long, yellow dashes on each side. The female's and male's face and eyes are colored similarly.

The Bleached Skimmer favors small, shallow, alkaline, desert pools and springs, especially hot springs. It can be found perched on alkaline flats and low brush near water or perched on low weeds. Bleached Skimmers oviposit in tandem, which is unusual for skimmers.

The Bleached Skimmer is a southwestern species ranging from Texas westward and north into southeastern Oregon. Its range extends southward into Mexico. In Oregon it is known only from alkaline lakes and hot springs (Borax Lake, Mickey and Alvord Hot Springs) in the Alvord Basin, Harney County at about 4,000 feet in elevation. It is common at these few locations.

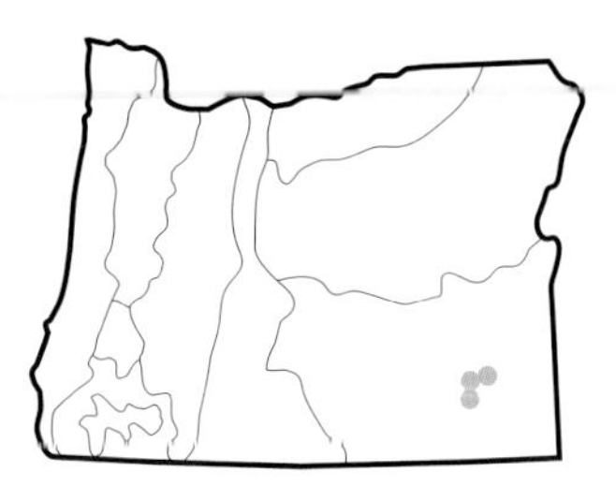

The adult flight season extends from mid-June to late August.

JAN	FEB	MAR	APR	MAY	JUN	JUL	AUG	SEP	OCT	NOV	DEC

Bleached Skimmer male (inset: side view)

Bleached Skimmer female (inset: side view)

Skimmers

Medium: 42–49 mm

Black and White Skimmer Identification

Identification Chart M illustrates male wing patterns for seven skimmer species which can pose field identification problems. We have included neither the distinctive Flame Skimmer with a large, reddish-orange body and wings, nor the two unique blue-bodied arid country skimmers: 1) Comanche with unique white in the pterostigma, and 2) the Bleached with unique pale white costa and orange tint at the wing base.

Of the remaining seven skimmer species, three are easy to separate with unique skimmer wing features: 1) both male and female Four-spotted Skimmers are dull and brown-bodied with a small dark spot at each nodus and brown streaking at the base, 2) both male and female Widow Skimmers show extensive bold, black patches on the basal half of each wing, and 3) the Hoary Skimmer has a dark smudge at the nodus and an elongated dark spot at the base; the male has some white pruinescence near the base, while the female has none.

Both whitetail species have flat, bright white abdomens, clear wing tips, and a wide dark band on the outer wing. Note the zig-zag, wavier edges on the Desert Whitetail's black marks. Also notice that the Desert Whitetail's extensive white pruinescence extends from the base to the black band, whereas the Common Whitetail shows a clear gap between the white base and that black mark.

The male Twelve-spotted Skimmer is unique among the male skimmers with a black tip on each long wing. Each wing has three black spots (at the base, nodus, and tip). The male Eight-spotted Skimmer shows black at the base and a "figure eight"-shaped black mark just beyond the nodus with white pruinescence beyond each black wing mark (in sequence from the base: black, white, black, white, and a clear wing tip).

The female skimmers generally have wing patterns similar to the males, but lack the white pruinescent marks. Hoary, Four-spotted, Eight-spotted, Widow, and Desert Whitetail are easy to match with the male patterns shown on the Identification Chart. Remember the unique zig-zag pattern on the female Desert Whitetail. The most difficult females to separate in the field are the Twelve-spotted Skimmer and the Common Whitetail, both with black wing tips. The female Common Whitetail is an exception from the general rule; unlike the male, the wing tip is black. The Twelve-spotted female has a yellow stripe alongside the abdomen, whereas the Common Whitetail female has a row of broken yellow dashes.

Widow Skimmer female

Identification Chart M • Black and White Skimmers, Genera: *Libellula* and *Plathemis*

Species	Notes (♂ wings)	♂ Wing
Eight-spotted Skimmer *Libellula forensis*	**2 large dark spots on each wing;** full "figure-8" band beyond nodus; basal dark streak; white beyond dark spots	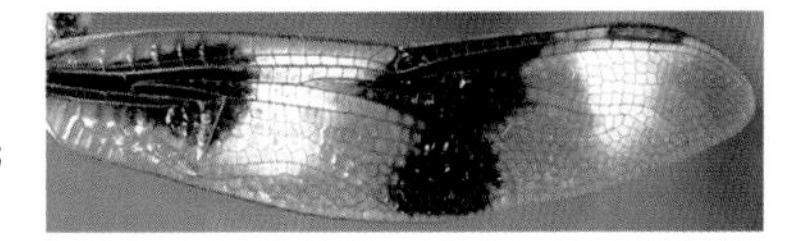
Widow Skimmer *L. luctuosa*	**Large basal dark spot (extends to nodus);** white beyond; wing tips clear	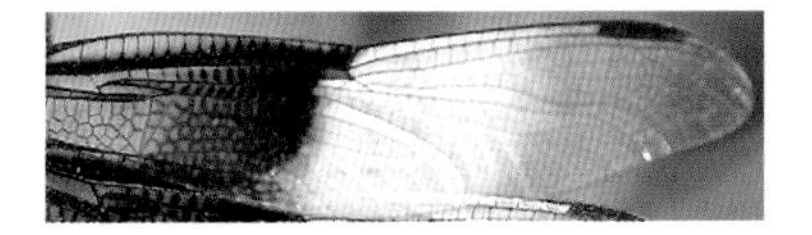
Hoary Skimmer *L. nodisticta*	**2 small dark spots on each wing;** small spot at each nodus; dark streak at base; white behind basal dark spot	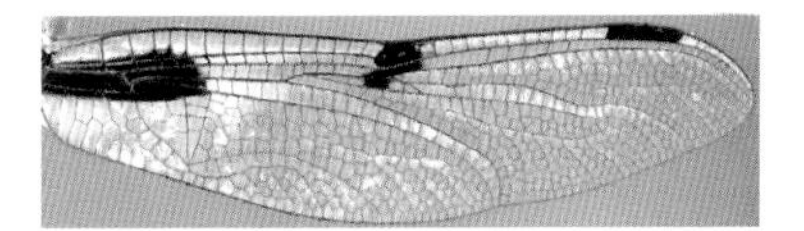
Twelve-spotted Skimmer *L. pulchella*	**3 dark spots on each wing;** base, mid-wing, and tip; white in between; wings very long	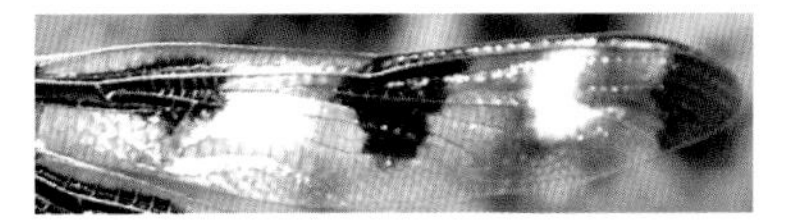
Four-spotted Skimmer *L. quadrimaculata*	**Wings mostly clear;** 4 small spots, 1 at each nodus; dark triangular spot at base of hindwing; amber wash on leading edge	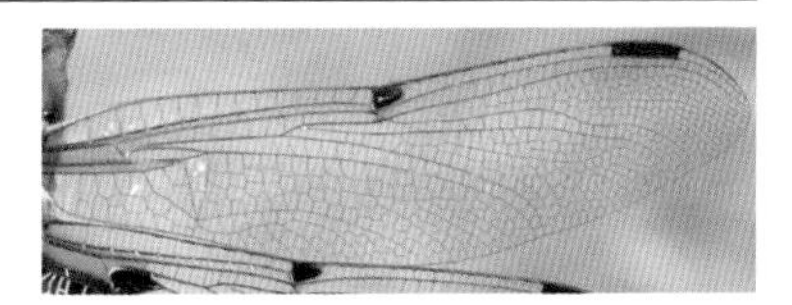
Common Whitetail *Plathemis lydia*	**Large mid-wing dark band;** basal dark streak; small white spot behind basal dark spot	
Desert Whitetail *P. subornata*	**Large mid-wing dark band;** dark basal streak; basal remainder extensive white	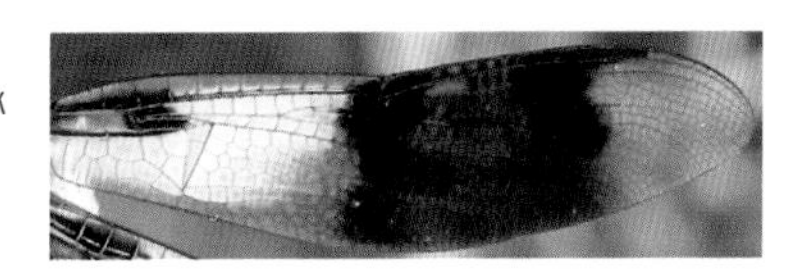

Eight-spotted Skimmer
Libellula forensis

The male shows an elongated, dark mark on the front of each wing base and a large, dark, "figure eight-shaped" mark just beyond the nodus with white patches beyond each dark mark. The dark basal mark on the wing is surrounded by white. The clear wingtips are important to note to compare to the similar Twelve-spotted Skimmer which has dark wingtips. The eight dark spots on the wing are the origin of the English name. The face and eyes are dark brown and almost metallic colored. The thorax and abdomen become covered in bluish-gray pruinosity. The female also shows two dark wing spots on each wing, but most females lack the white patches or the patches are fainter than the male's. The female has a dark brown abdomen with broken yellow stripes on the side.

The Eight-spotted Skimmer can be found at a wide range of habitat types, including ponds, lakes, marshes, ditches, and slow waters along rivers and streams that have emergent vegetation. Males often perch atop stems or plants and fly over water to hunt or fly long patrols before returning to the same perch. Males chase other males, including males of other species.

The Eight-spotted Skimmer is a western species ranging from Mexico north into southernmost British Columbia. It is found throughout Oregon: primarily below 1,000 feet in elevation in the northern and western regions, up to at least 6,000 feet in southern and southeastern regions. It is one of our most common urban species and is appreciated for its bright wing pattern. It has been recorded in all thirty-six Oregon counties.

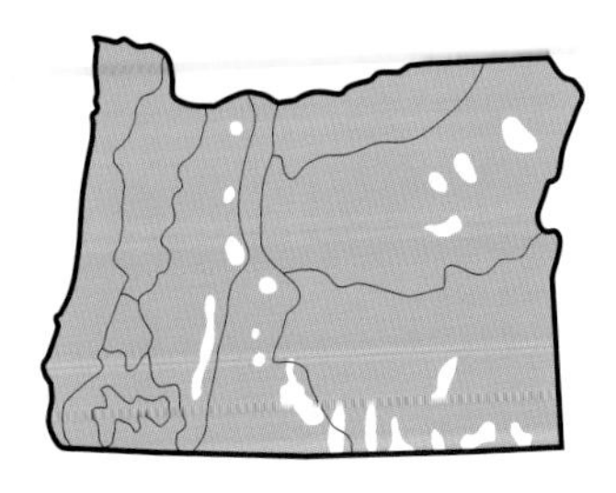

The adult flight season extends from early May to late October.

JAN	FEB	MAR	APR	MAY	JUN	JUL	AUG	SEP	OCT	NOV	DEC

Eight-spotted Skimmer male

Eight-spotted Skimmer female

Medium: 44–51 mm

Widow Skimmer

Libellula luctuosa

The attractive Widow Skimmer shows a distinctive and unmistakable wing pattern. The mournful, black, cloak-like wing base is the source of its name. The male has a black basal patch on each wing that extends almost all the way to the nodus. In mature males, this basal patch is bordered by white pruinescence to the pterostigma. The female's basal patch is brown, not as intense as the male's, lacks the white bordering mark, and shows a light brown wash at the wingtip. The male's face, eyes, and thorax are dark brown, with pruinose gray on the top of the thorax and a pruinose bluish-white abdomen with a dark tip. The female has a brown thorax with large tan spots low on the sides and a broad, brown abdomen, which has a dark medial stripe on top, bordered on each side by a yellow stripe. The yellow stripe on the female starts on the thorax, and splits at the base of the abdomen into two stripes. The wing patterns of both male and female are unique, making field identification easy.

Widow Skimmers can be found at the vegetated shores of ponds, lakes, marshes, and other still waters. They fly in a seemingly unsteady manner low over water and often perch on the tips of reeds or twigs at the water's edge. Widow Skimmers have adapted well to constructed ponds and waterways, which may have led to a century of range expansion; they were not known in California until the early 1900s (Manolis, 2003). Widow Skimmers may wander far from water into parks, yards, and gardens.

The Widow Skimmer is transcontinental, and in the west it ranges from Mexico north into Washington. In Oregon, it is found in western lowlands from California to Washington below 1,000 feet in elevation. It is expanding its range northward, and was first recorded in Oregon in 1991.

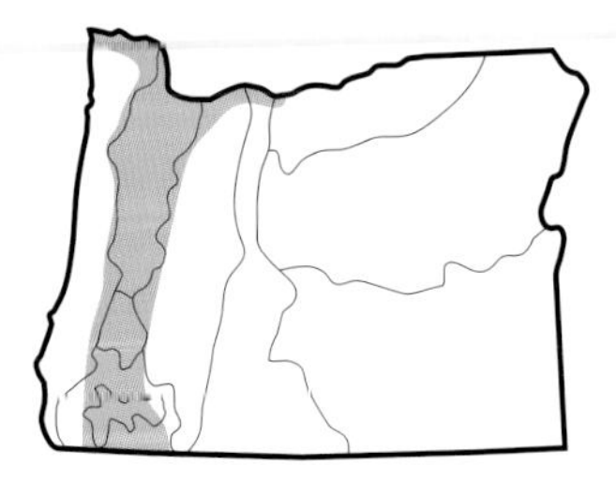

The adult flight season extends from early June to mid-October.

JAN FEB MAR APR MAY JUN JUL AUG SEP OCT NOV DEC

Widow Skimmer male

Widow Skimmer female

Medium: 44–50 mm

Hoary Skimmer

Libellula nodisticta

The mature Hoary Skimmer male is covered with pale, blue-gray pruinosity, which obscures much of the underlying brown and yellow color. The wing pattern is unique. Each wing has an elongated dark patch at the base, a smaller patch at the nodus, and a dark pterostigma. The dark, nodal patch is the source of its scientific name, and the white pruinosity looks like hoarfrost, thus its English name. At maturity the basal patches become surrounded by bluish-white pruinosity. Underlying the pruinose body, the thorax is brown with two pairs of yellow spots on each side. The face is yellowish and the eyes and legs are black. The female is grayish-brown with a row of yellow dashes along the sides of the abdomen. The female has a yellowish-brown face, and her thorax is brown with two yellow spots on the side. The female and male wing patterns are similar. Both sexes have a very hairy thorax. At an advanced age, some females become pruinose like the male.

The Hoary Skimmer is found at elevations from 4,000 to 5,000 feet in cold and warm springs, small streams, and *Darlingtonia* fens in the Siskiyou Mountains. It perches atop sedges and weeds near water and forages well into dusk (Needham, et. al., 2000). Sometimes, pairs fly in tandem while ovipositing. Other than the Bleached Skimmer, this is a rare behavior among skimmers.

The Hoary Skimmer is a southwestern species ranging from the Oklahoma panhandle westward and from Mexico northward into southern Oregon and Idaho. It is most common in the far southeastern corner of Oregon in Lake, Harney, and Malheur Counties with a few records from southern Klamath and Josephine Counties.

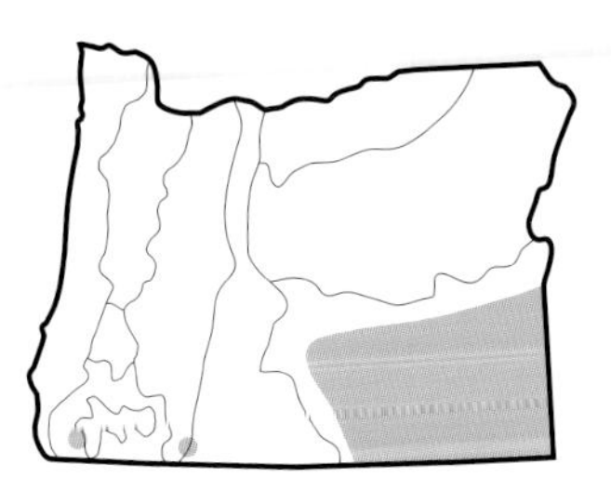

The adult flight season ranges from late May to late August.

JAN FEB MAR APR MAY JUN JUL AUG SEP OCT NOV DEC

Hoary Skimmer male

Hoary Skimmer female (inset: pruinose female)

Medium: 46–52 mm

Twelve-spotted Skimmer
Libellula pulchella

The scientific name, *pulchella,* refers to the beautiful color pattern on the mature male. Each wing shows three large, dark spots giving this large species its English name, "Twelve-spotted." With one dark spot at the tip, nodus, and base of each wing, white spots between each dark spot, and an additional white spot near the base of each hindwing, the male is easily differentiated from our other king skimmers. The male becomes thinly bluish-gray pruinose over the thorax and abdomen. Underlying the pruinosity is a brown and gray thorax and abdomen patterned similarly to the female's body. The face and eyes are shiny black. The forewings of the Twelve-spotted Skimmer are longer than those of the other king skimmers, a feature that is noticeable in the field. The female's wings also have three spots each, but lack any white pruinosity. The female shows continuous, unbroken yellow stripes on each side at the top of the brown abdomen. The female Common Whitetail also shows three dark wing spots, including the dark wing tip, but has a series of broken yellow spots along the side of the abdomen.

The Twelve-spotted Skimmer is common at ponds, lakes, and other slow waters with emergent vegetation. It often perches near shore in sunny areas on tall weeds and stems, and males aggressively defend their territory from other males, including those of other species. The males often sally over the water and return to a favorite perch. This dragonfly is a strong flier and can perform amazing acrobatic aerial maneuvers.

The Twelve-spotted Skimmer's range is transcontinental, ranging from Mexico north into southern Canada. It ranges widely throughout Oregon—primarily below 1,500 feet in elevation in northern and western Oregon and up to 7,000-foot elevation in the southern Cascades and Basin and Range.

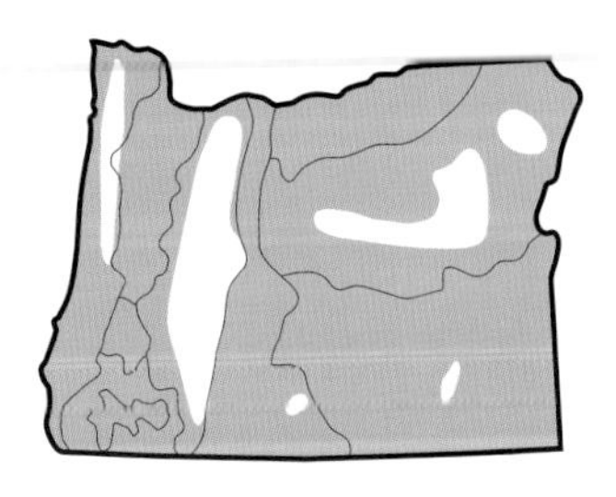

The adult flight season extends from late May to early October.

JAN	FEB	MAR	APR	MAY	JUN	JUL	AUG	SEP	OCT	NOV	DEC

Twelve-spotted Skimmer male

Twelve-spotted Skimmer female

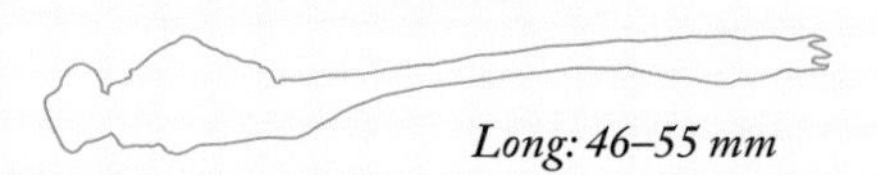

Long: 46–55 mm

Four-spotted Skimmer

Libellula quadrimaculata

This common, very hairy dragonfly is olive-brown and nondescript overall. It shows yellow stripes on the side of the thorax and a series of long, yellow spots along the sides of the top of the abdomen. The abdomen is stubby and tapered, with a black tip. When mature, these spots are very subtle. Tim Manolis aptly described the Four-spotted Skimmer abdomen from above as looking like it was dipped in ink (Manolis, 2003). Each wing has a dark spot at the nodus, and the hindwings have a triangular black spot at the base with an orange to brown wash at the front. These spots are the source of its English name "Four-spotted." The fronts of the wings also have a golden wash in the veins from the base to near the nodus, and the pterostigma are dark. The head and eyes are dark brown. The female and male are similar, a trait unusual among the skimmers. The female's abdomen is much broader with shorter cerci. The overall impression is of a very stout and sturdy dragonfly.

The Four-spotted Skimmer is found at still waters at lakes, marshes, and bogs, with plentiful vegetation and muddy bottoms. At some places it can be abundant, as it perches on low, emergent vegetation or on the ground. The males are aggressive and frequently chase other males. As the females oviposit, males guard them by hovering nearby.

The Four-spotted Skimmer is found in northern latitudes throughout North America, Europe, Asia and as far south as Northern Africa. It is very common at higher elevations (up to 7,000 feet) throughout Oregon, and it is scarce at lowland elevations.

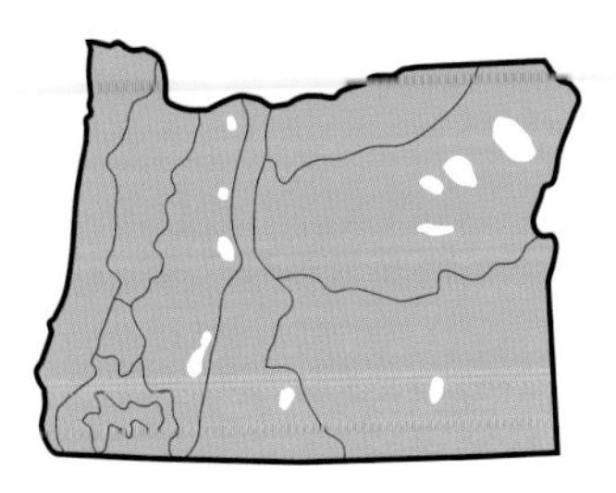

The adult flight season ranges from mid-May to mid-September.

JAN	FEB	MAR	APR	MAY	JUN	JUL	AUG	SEP	OCT	NOV	DEC

Four-spotted Skimmer male

Four-spotted Skimmer female

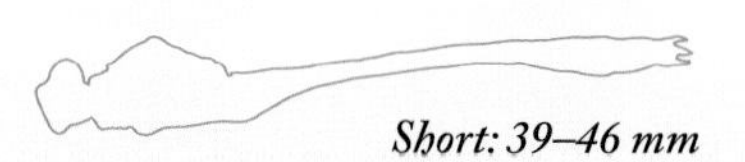

Short: 39–46 mm

Common Whitetail

Plathemis lydia

The male Common Whitetail has a stout, much-flattened abdomen. The abdomen is covered in thick, bright white pruinosity at maturity, giving this species its English name, "whitetail." The head, eyes, and thorax are dark brown. Each wing has two black patches, an elongated streak at the base behind the costa and a broad, full band at the nodus with clear spots on the wingtips. This combination of wing pattern and bright white abdomen makes the male easy to differentiate from other king skimmers in the field. The female has three dark spots on each wing including one at the wing tip. The brown abdomen of both the immature male and the female have a series of oblique yellow spots along the sides of the top of the abdomen. With three dark spots in each wing, the female can be difficult to differentiate from the female Twelve-spotted Skimmer, which has an unbroken yellow side line on the abdomen, in contrast to the female Common Whitetail's series of broken yellow spots.

Males and females frequently perch on the ground—or other light colored objects close to the ground, such as flat rocks, broken branches, and occasionally low vegetation. From its low perch, the Common Whitetail launches itself to catch insects that fly by. The males aggressively defend shoreline territories and chase other males, including males of other similar species. Males hover-guard females during oviposition. Females lay eggs in sheltered places near shore, often ovipositing several times in one location. Females may lay 25–50 eggs at each tap of the abdomen and may lay around 1,000 eggs total (Paulson, 2009).

Common Whitetails can be found at ponds, lakes, marshes, roadside ditches, sloughs, quiet pools, and other slow waters, including beaver ponds. It wanders into parks, yards, and gardens. The Common Whitetail is transcontinental, ranging from southern Canada into Mexico. In Oregon it is common at lowland ponds and lakes below 1,000 feet in elevation. In the southern Cascades and Basin and Range regions it is found up to at least 5,000 feet, but it is scarce at those higher elevations.

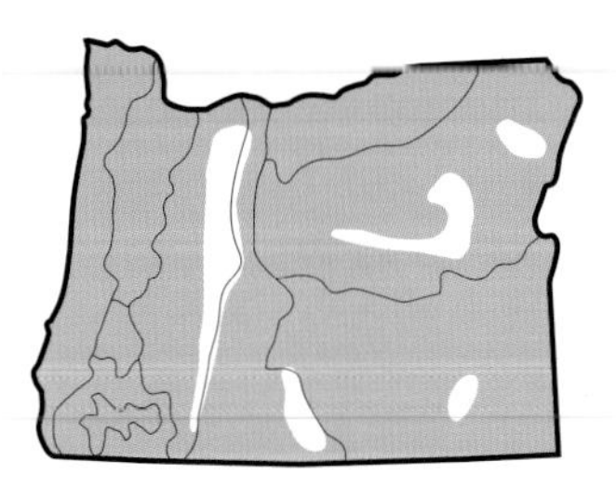

The adult flight season extends from early May to early October.

JAN	FEB	MAR	APR	MAY	JUN	JUL	AUG	SEP	OCT	NOV	DEC

Common Whitetail male

Common Whitetail female

Medium: 42–48 mm

Desert Whitetail

Plathemis subornata

This lovely species' scientific name, *subornata*, refers to ornate, yellow spots on the underside of the abdomen. The vivid, mature, male Desert Whitetail shows white pruinosity like the Common Whitetail, but the Desert Whitetail has a bold, unmistakable wing pattern. It has a broad black band with jagged edges that extends from the nodus to the pterostigma and a smaller black mark at the base near the front of each wing. The area between the two black marks is white and the wingtip is clear. The male has dark brown eyes and a dark face with yellow spots on the upper side. The female's thorax is brown with broad, yellow, broken side stripes. The female's face is yellow. The immature male's and female's wings are marked similarly with the broad, black band broken into two zig-zag dark marks by an intervening clear area. As males mature, the clear spot fills in with brown turning to black. Nevertheless, the wing pattern of both sexes is unique for field identification. Both the female and immature male have a brown abdomen with a row of bold, yellow dashes on either side of a brown median stripe.

Desert Whitetails frequent springs and pool edges in arid country. They fly low over the ground before perching low on brush and nearby vegetation, commonly sagebrush. They often perch on sedges and other low vegetation at the water's edge. In Oregon, the Desert Whitetail has only been found at wetlands associated with alkaline hot springs.

The Desert Whitetail is a southwestern species with a range extending westward from western Kansas and Oklahoma, and from Mexico north into southern Oregon and Idaho. In Oregon it is known only from the Basin and Range region in Malheur, Harney, and Lake Counties from 4,000 to 5,000 feet in elevation.

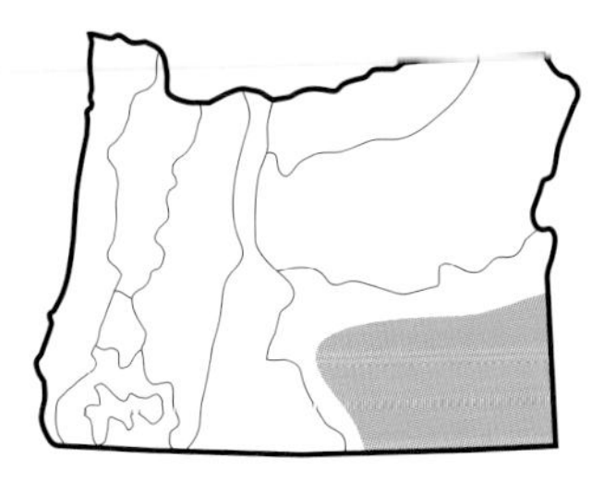

The adult flight season extends from mid-May to late August.

JAN	FEB	MAR	APR	MAY	JUN	JUL	AUG	SEP	OCT	NOV	DEC

Desert Whitetail male (inset: immature male)

Photo by Steve Gordon

Desert Whitetail female

Medium: 37–51 mm

Flame Skimmer

Libellula saturata

The gorgeous Flame Skimmer male is large and unmistakably red-orange; its scientific and English names refer to being saturated in flame colors. It has a reddish-orange face, eyes, unmarked brownish-red thorax, and reddish-orange abdomen, legs, and wing veins. The wings show an orange wash extending from the base to beyond the nodus on each wing with brown streaks at the base, orange on the leading edge of each wing, and an orange pterostigma bordered in black. The female and immature male are more tan, but show a pale, yellowish median stripe atop the thorax extending between the wing bases and a more washed-out red basal patch in the wing. The female has a tan face with red-brown eyes.

Males patrol long stretches of streams and shorelines, and perch at eye-level in weeds over water or in nearby brush and trees. They feed by sallying out from a perch to catch insects, and males are very territorial. They mate in flight and the female flips eggs into the water, often spreading them to the water line or onto shore (Manolis, 2003).

In western Oregon it is common at ponds, lakes, ditches, and slow moving streams. East of the Cascades it is found at non-alkaline warm springs and some sluggish streams. Flame Skimmers wander far from water and often show up in parks, yards, and gardens.

The Flame Skimmer is a western species extending westward from eastern Texas and northward from Mexico to the Canadian border. In Oregon, it is known west of the Cascades up to 1,500 feet elevation. It is common in the Rogue, Umpqua, and southern Willamette Valleys, and has been found on the south coast. It appears to be expanding its range northward. East of the Cascades it is locally common in the Basin and Range and Blue Mountain regions.

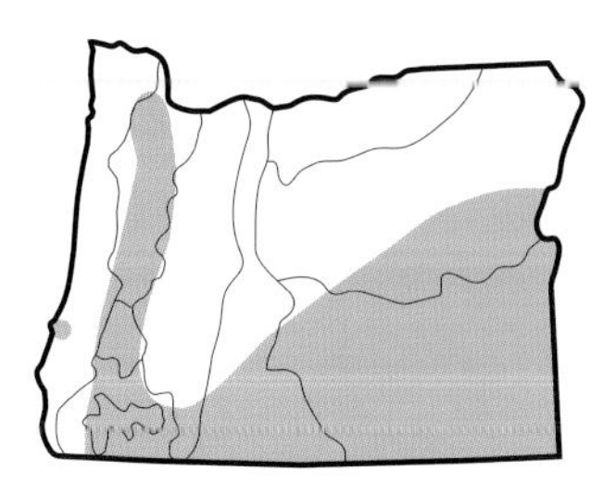

The adult flight season extends from late May to early October.

JAN	FEB	MAR	APR	MAY	JUN	JUL	AUG	SEP	OCT	NOV	DEC

Flame Skimmer male

Flame Skimmer female (with egg mass)

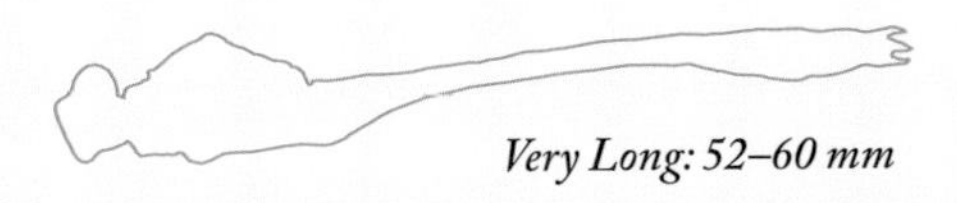

Very Long: 52–60 mm

Red Rock Skimmer

Paltothemis lineatipes

The male Red Rock Skimmer is easily identified by its red and black color. The face and eyes are red. The thorax and abdomen are intricately patterned in subdued red and black. The long, broad wings have an amber wash at the base which extends to just short of the nodus with red veins in the colored area. The size of the wings is impressive compared to the size of the body. The female is much duller colored than the male. The eyes are dark brown, and the face is pale. The pattern on the thorax and abdomen is pale brown and black. The female lacks the color at the base of the wings. Our other similarly-colored red dragonflies like the Cardinal Meadowhawk and Flame Skimmer are much brighter and lack the intricate black markings on the thorax and abdomen.

The Red Rock Skimmer frequents rocky streams where it perches on emergent rocks or on rocks along the shoreline in sunny locations. When disturbed, it will often fly a short distance and perch again. During copulation, males display oviposition sites to the female by flying back and forth over them (Corbet, 1999). Males hover guard females during oviposition, but females may also oviposit alone (Manolis, 2003).

This skimmer is a species of the arid southwest ranging from southern Oregon to Utah and south into Mexico (Needham, et. al., 2000). In Oregon, the northernmost location, it is only known from Cottonwood Creek near Fields in the Alvord Desert. However, it has not been found there in several years, so its status in Oregon is unknown. It should be sought at similar small, rocky streams in Oregon's desert country.

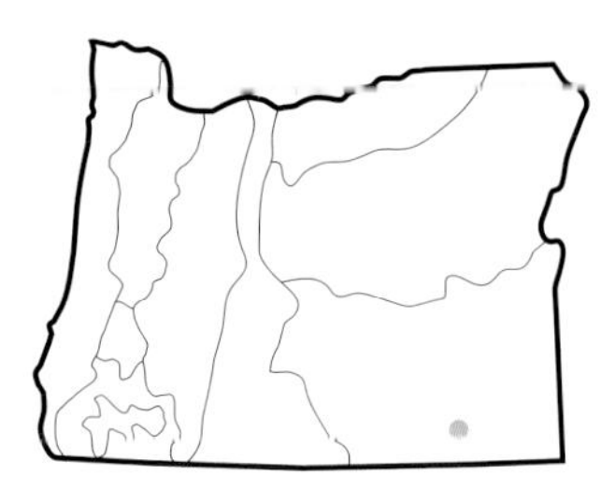

The adult flight season is from mid-June to late August.

JAN	FEB	MAR	APR	MAY	JUN	JUL	AUG	SEP	OCT	NOV	DEC

Red Rock Skimmer male

Red Rock Skimmer female • photo by Dennis Paulson

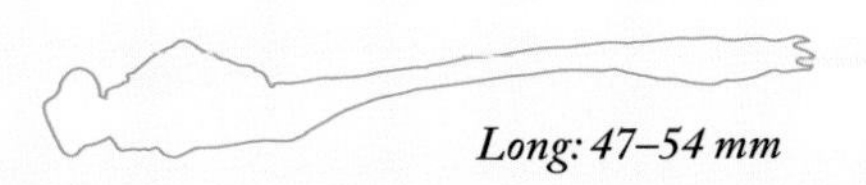

Long: 47–54 mm

Variegated Meadowhawk in flight

Cardinal Meadowhawk male obelisking

Meadowhawks

The ten meadowhawks can be difficult for beginners to identify to species. Refer to Identification Chart N for key meadowhawk features along with male side views. These characters will also assist with identification of the females. The meadowhawks spend a lot of time perching on vegetation, the ground, and rocks and can be approached fairly closely. While some of the characters in the chart are difficult to see, this behavior makes it easier to get close enough to see characters with binoculars.

The first five species listed in the table are easily identified due to their distinctive thoracic pattern. The Black Meadowhawk is unique with a black and yellow patterned thorax, and is the only meadowhawk with this black color. The Variegated Meadowhawk has a distinct abdominal color pattern with white spots on the side surrounded by black on a red background.

The Cardinal Meadowhawk is bright red with two yellow or white spots on the side of the thorax and an unmarked abdomen, while the Striped Meadowhawk is duller in color with light stripes on the sides of the thorax and black markings on the abdominal sides. The Band-winged is the only meadowhawk with the basal half of the wing colored. It has complex markings on the side of the thorax and dark spots atop S8 and S9.

The next five meadowhawks have plain thoracic sides. The Cherry-faced and White-faced Meadowhawks are quite similar with the exception of face color. There is variation in the color of the face on the Cherry-faced, but the face will not be white as in the White-faced.

The next two species, the Red-veined and Saffron-winged Meadowhawks, are difficult to identify as the color in the leading edge of the wing is variable. Check the colors on the leg to aid in identification of these two species: all black in the Red-veined and black with brown tibial stripes in the Saffron-winged.

The final species in the chart, the Autumn Meadowhawk is a late-season species and can be identified from its uniform body color, very small size, and thin, unmarked abdomen. The female also has a unique spout-like subgenital plate that can be seen in side view.

Identification Chart N • Meadowhawks, Genus *Sympetrum*

Species (size)	Head	Thorax	Wings	Abdomen	Legs	Photo
Black *Sympetrum danae* Very Short 27-35 mm	All black	Black, may show yellow	Clear	Black	Black	
Variegated *S. corruptum* Long 40-42 mm	Red face; brown eyes	Faint stripes ending in yellow spots	Red veins on leading edge	White spots bordered in black	Black with brown stripes	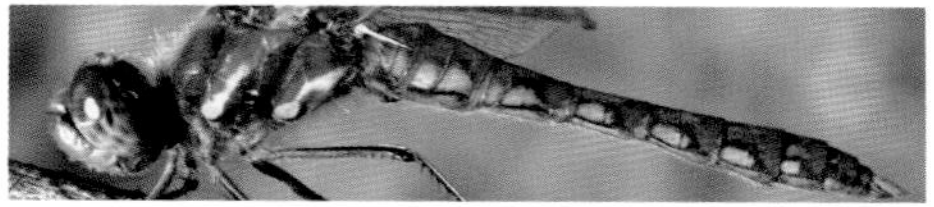
Cardinal *S. illotum* Medium 37-40 mm	Bright red	Two white or yellow spots	Red in front veins	Bright red, unmarked	Brown	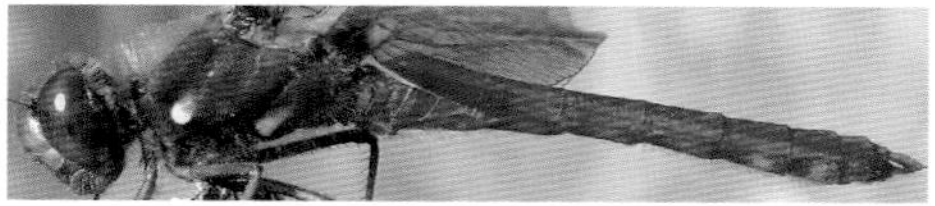
Striped *S. pallipes* Medium 32-36 mm	Pale face; red-brown eyes	Two side stripes	Clear	Small black markings on lower sides	Black	
Band-winged *S. semicinctum* Medium 30-40 mm	Reddish face and eyes	Complex markings	Orange, base to nodus	Black along lower side	Black	

Identification Chart N • Meadowhawks, Genus *Sympetrum* (continued)

Species (size)	Head	Thorax	Wings	Abdomen	Legs	Photo
Cherry-faced ***S. internum*** Short 31-36 mm	Orange face, brown eyes	Plain	Amber wash at base, orange veins	Black saw-toothed pattern	Black	
White-faced ***S. obtrusum*** Short 30-39 mm	White face	Plain	Clear	Black saw-toothed pattern	Black	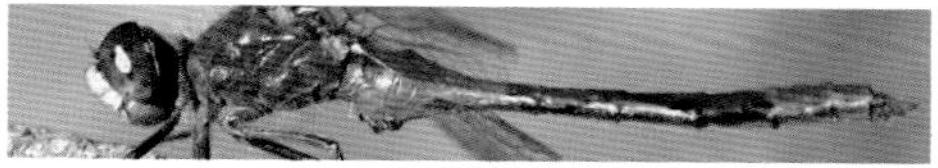
Red-veined ***S. madidum*** Long 37-45 mm	Red face and eyes	Plain	Red or amber veins on leading edge	Black saw-toothed pattern	Black	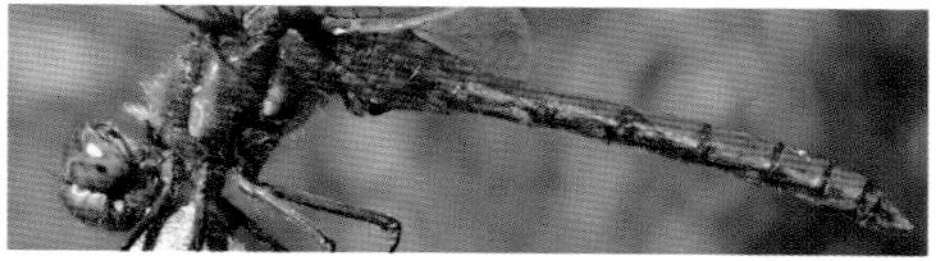
Saffron-winged ***S. costiferum*** Medium 31-38 mm	Red-brown face and eyes	Plain	Golden veins on leading edge	Black on lower sides	Black, brown stripes on upper sides	
Autumn ***S. vicinum*** Short 31-37 mm	Red-brown face and eyes	Plain	Clear	Small, thin abdomen	Brown	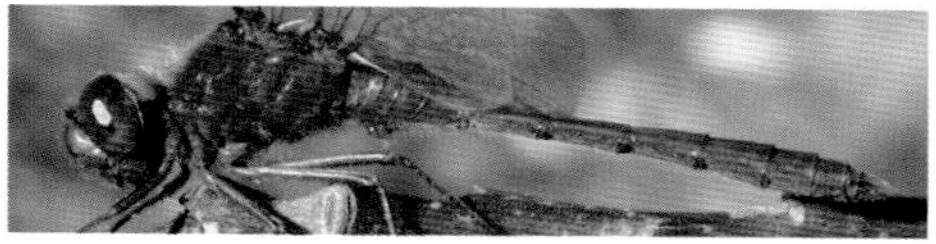

Variegated Meadowhawk

Sympetrum corruptum

This species is named after its complex color pattern. The Variegated Meadowhawk is a robust species with thoracic stripes showing as faint white stripes ending in yellow spots on the lower end. The thorax is brown with some variation, but the low, yellow spots are always visible when white stripes may not be apparent. The veins on the leading edge of the wings are red. The eyes are dark brown on top, and the face is red. The abdomen has a complex pattern of white bordered by black, with red in the intervening areas. The legs are black with yellow and orange stripes. The female thorax appears more gray but still has the faint white lines on the side ending in yellow spots. The face is pale, and the eyes are gray. The same characteristic white spots bordered in black are found on the sides of the abdomen. The top of the abdomen is gray with orange segment borders. The veins on the leading edge of the wings are yellow. The Variegated Meadowhawk's white abdominal side spots bordered by black are diagnostic and set it apart from our other meadowhawks. In the field, look for these spots on the side of the abdomen for identification.

The Variegated Meadowhawk is widespread across the United States and into southern Canada extending as far north as central Alberta. This meadowhawk species is very common throughout Oregon up to at least 7,300 feet in elevation at lakes, marshes, ponds, and even in urban areas. This is one of our hardiest species. It is sometimes observed migrating south along the Oregon Coast in the fall in large numbers, and the earliest observations in the spring are likely from migrations from the south.

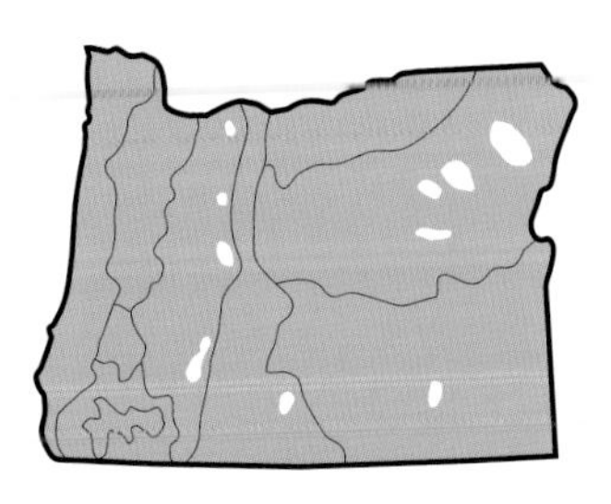

The adult flight season extends from early January to late October.

JAN	FEB	MAR	APR	MAY	JUN	JUL	AUG	SEP	OCT	NOV	DEC

Variegated Meadowhawk male

Variegated Meadowhawk female

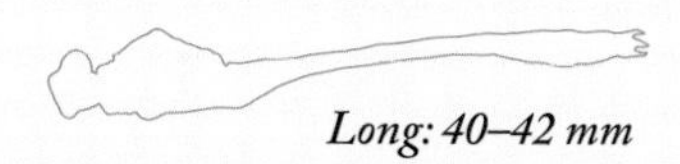

Long: 40–42 mm

Saffron-winged Meadowhawk

Sympetrum costiferum

Both the scientific name, *costiferum*, and the English name, "Saffron-winged," refer to the costal stripe on the wing's leading edge. The thorax of the Saffron-winged Meadowhawk is plain and unmarked deep red to reddish-brown while the wings show golden or yellowish veins forming a distinctive stripe at the leading edge. The thorax may show white stripes in immature males. This feature can be similar to the Red-veined Meadowhawk but note the smaller size of the Saffron-winged Meadowhawk. The pterostigma is reddish with dark borders. The abdomen is red with the underside black, and there is variable black striping on the center of the top of segments 7 to 9. The eyes are two-toned, red above and yellow below. The legs are black with light brown on the upper side of the femur and tibia; in contrast, the legs of the Red-veined Meadowhawk are always black. The female has a thorax that is tan on top, yellow with small black markings on the sides, and wings that have yellow veins forming a stripe on the leading edge. The pterostigma is yellow with black borders. The face is yellow-green, and the eyes are dark reddish-brown on top blending to yellow on the bottom. The yellowish abdomen has a black stripe on the bottom as well as some mottled black on the lower sides.

The Saffron-winged Meadowhawk is found in Southern Canada and the northern and central U. S. with scattered occurrences from the Canadian Northwest Territories to New Mexico. It is found across Oregon and throughout the Willamette Valley at vegetated ponds, lakes, and slow-moving streams. It is found up to 6,100 feet in elevation in the southern Cascades, but is absent at higher elevations elsewhere. In Lane County, we have observed this uncommon species most often in grassy areas along Willow Creek, at Sandpiper Pond, and at Fern Ridge, which are all in the west Eugene wetlands.

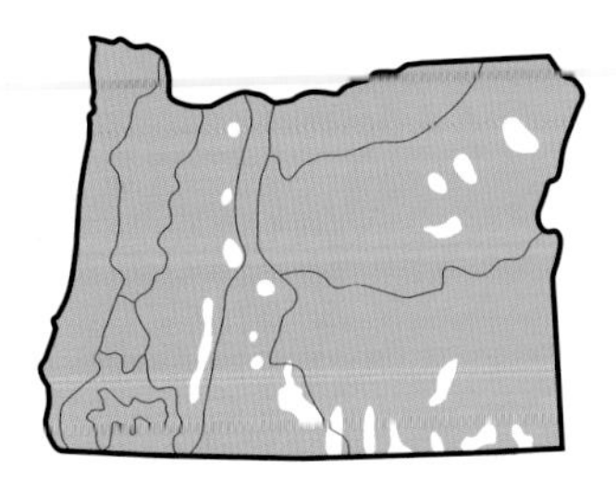

The adult flight season extends from early July to early October.

JAN	FEB	MAR	APR	MAY	JUN	JUL	AUG	SEP	OCT	NOV	DEC
						▬	▬	▬			

Saffron-winged Meadowhawk male

Saffron-winged Meadowhawk female • photo by Dennis Paulson

Medium: 31–38 mm

Black Meadowhawk

Sympetrum danae

This small, dark meadowhawk is unlike all of our other species. The head, face, and thorax are brownish-black. The legs and pterostigma are black as is the abdomen. The abdomen may have some paired lighter areas on the top. The female has a yellowish face, brown top of the thorax, and yellow and black on the sides of the thorax. The female's abdomen is yellow-brown on the top and black on the sides and bottom. The female may have an amber wash at the base of the wings. The immature male is colored like the female. Because of the color and small size, this species is easily identified in the field.

The Black Meadowhawk is found across the northern part of the United States into the far north of Canada and Alaska. In the west, it ranges as far south as central California. It is found in Oregon's mountainous regions up to 7,500 feet in elevation and at lower elevations in eastern Oregon.

It is very common at marshes and wet, boggy ponds in the mountains such as Gold Lake Bog. Here, it will be commonly seen perched on emergent vegetation over or near water. Females and immature males are usually found perched in nearby vegetation. Females oviposit in tandem or singly by tapping the abdomen to water, moss, sometimes mud, and may even drop eggs from the air (Paulson, 2009).

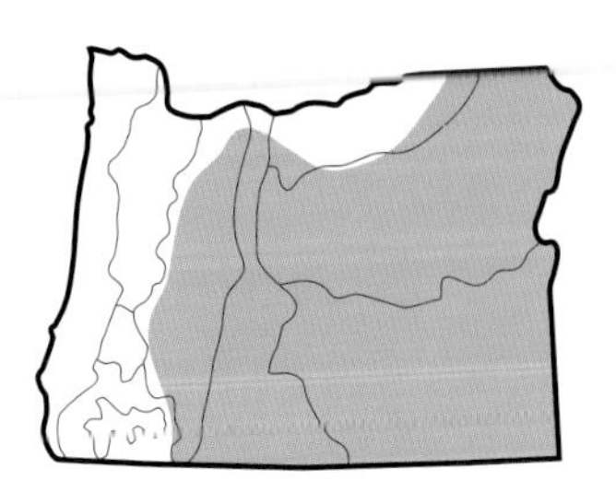

The adult flight season extends from late June to early October.

JAN FEB MAR APR MAY JUN JUL AUG SEP OCT NOV DEC

Black Meadowhawk male (inset: male, top view)

Black Meadowhawk female

Very Short: 27–35 mm

Cardinal Meadowhawk

Sympetrum illotum

This species stands out among the meadowhawks due to its bright, cardinal red color. The side of the thorax has two faint stripes that often show as white spots on the lower end. The wings have orange in the area around the front two or three veins from the base to the nodus. The eyes and face are red and the abdomen is red and unmarked. This common meadowhawk appears almost pure bright red and can be told from the Flame Skimmer by its smaller size and much narrower abdomen. However, the Cardinal Meadowhawk's abdomen appears thicker than our other meadowhawks. The female appears browner than the male in color, with dark brown eyes and a lighter face. The sides of the thorax have white spots similar to the male. The female's abdomen is brown with black at the seam that runs along the sides. The legs of both sexes are light to dark brown. The female and male wings are similar, with a reddish-brown pterostigma.

In North America, the Cardinal Meadowhawk is found on the west coast from southern British Columbia into the southwest. It is common throughout lowlands west of the Cascades but is less common on the east side. In the field, you will note this small, bright red dragonfly perched over the water with its wings flexed downward. On hot days, it becomes an acrobat—perching with the abdomen pointed towards the sun to reduce the surface area and maintain a cooler temperature. It can be approached and will usually return to the same perch site over and over again. It is found throughout the Willamette Valley near ponds and in nearby grassy areas even in the urban environment.

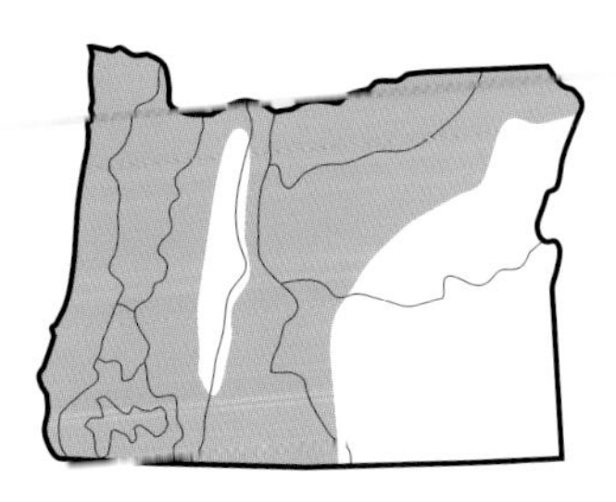

The adult flight season extends from late March to early November.

JAN	FEB	MAR	APR	MAY	JUN	JUL	AUG	SEP	OCT	NOV	DEC

Cardinal Meadowhawk male (inset: male, top view)

Cardinal Meadowhawk female

Medium: 37–40 mm

Cherry-faced Meadowhawk

Sympetrum internum

The overall impression of this Meadowhawk in the field brings into focus its red color and small size. The thorax is reddish-brown without markings, and the legs are black. The mature male's eyes are dark red and its face color is variable from yellowish-brown to red. The immature male's face is yellowish and slowly turns darker with time. The wings have a small amber wash at the base and orange veins. The abdomen is red with distinctive black markings on the side, which are large, horizontal triangles that appear in a saw-toothed pattern.

The Cherry-faced Meadowhawk can be distinguished from similar meadowhawks as follows: it has the black triangle pattern on the abdomen and black legs while the Saffron-winged Meadowhawk has narrow, black abdominal markings and black streaking on paler legs. Of Oregon meadowhawks, the Cherry-faced and White-faced Meadowhawks are similar in body color and pattern but can be separated by the obvious difference in face color. The female Cherry-faced Meadowhawk has an overall yellowish-brown appearance. The thorax is darker on top blending to yellow-green on the lower sides. The eyes are dark brown on top and yellow-green towards the bottom. The face is yellow-green, legs black, and the wings have an amber wash from the base to the nodus. There is a thick, black line marking the sides of the abdomen.

The Cherry-faced Meadowhawk is found across North America from the central U. S. to Alaska and the Yukon Territory. This meadowhawk lives between 2,500 to 5,800 feet in elevation in ponds, slow streams, and marshy meadows that are seasonally wet. Its distribution in Oregon is at lower elevations in eastern Oregon but there are a few records from Multnomah County as well as Eugene in Lane County. These individuals may be wanderers from the east, but small populations breeding in the Willamette Valley are not out of the question.

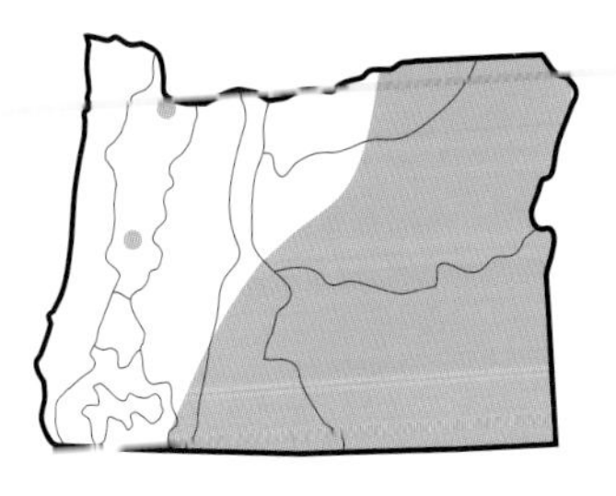

The adult flight season extends from early July to mid-September.

JAN	FEB	MAR	APR	MAY	JUN	JUL	AUG	SEP	OCT	NOV	DEC

Cherry-faced Meadowhawk male

Cherry-faced Meadowhawk female • photo by Ken Tennessen

Short: 31–36 mm

Red-veined Meadowhawk

Sympetrum madidum

The Red-veined Meadowhawk is similar to the shorter Saffron-winged Meadowhawk. The veins in the leading edge of the wing are red to amber. The markings on the dark red-brown thorax show as dots at the lower end of the light stripes, but may not be present in mature males. The face is red, eyes dark red, and the legs are black. The abdomen is dark red with two black lines along the sides and some pruinosity on the bottom. The overall color of the female is tan to brown, and the leading edge of the wing has tan to amber colored veins. The eyes, thorax, and top of the abdomen are tan. The thorax has two light-colored stripes on the sides, and the abdomen has two dark brown to black lines on the sides. The bottoms of the basal abdominal segments of mature females are pruinose.

The Red-veined Meadowhawk is found from southern California to Minnesota and in Canada into the Northwest Territories. It occurs across Oregon up to 4,500 feet in elevation, although currently it has not been found in some eastern Oregon counties.

This meadowhawk oviposits in pairs or alone in shallow waters which dry up or in dry basins—with eggs hatching after areas are refilled by winter rains (Paulson, 2009). This uncommon to locally common species frequents stagnant, seasonal ponds like shallow beaver ponds, and ponded areas in seasonal streams throughout the Willamette Valley. It is locally common at some wetlands in Oregon Coastal Dunes.

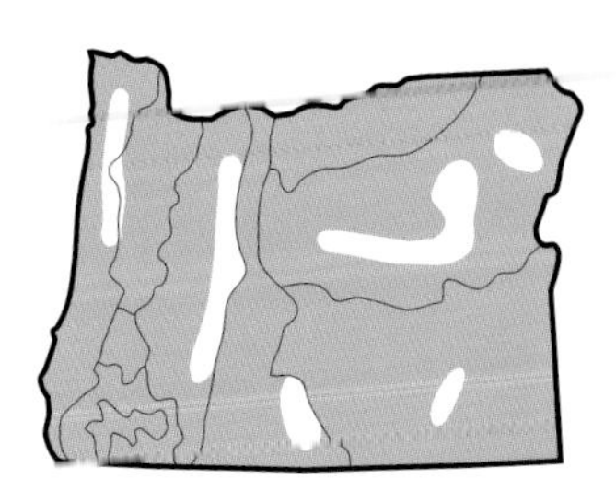

The adult flight season extends from mid-May to mid-September.

JAN	FEB	MAR	APR	MAY	JUN	JUL	AUG	SEP	OCT	NOV	DEC

Red-veined Meadowhawk male

Red-veined Meadowhawk female (inset: female side view)

Long: 37–45 mm

White-faced Meadowhawk

Sympetrum obtrusum

The mature male White-faced Meadowhawk has an unmarked brown thorax with some red areas on the sides and a slender, red abdomen with sawtooth-shaped black markings on the lower sides. The wings are mostly clear, but may show a touch of amber at the base, with orange wing veins along the front of the wing displaying a brown pterostigma. The face is white, as its name implies. The eyes are red, and the legs are black.

The White-faced Meadowhawk is similar to the Cherry-faced Meadowhawk. The major differences are the obvious white face in the White-faced and cherry-red face in the Cherry-faced. The veins of the leading edge of the wing are orange in the White-faced versus black in the Cherry-faced. The female White-faced is brown with an unmarked thorax, deep red-brown eyes, and also has a white face. The female, like the male, has black legs, and the brown abdomen has black sawtoothed markings similar to the male.

This meadowhawk is found across the northern and central U. S. and as far north as the Northwest Territories of Canada. It is common in Oregon's mountain ponds and lakes, especially those with adjacent wet meadows above 3,200 feet in elevation, including both permanent and seasonal waters (Paulson, 2009).

Females oviposit in tandem or alone guarded by the male. Thoracic temperature of the male determines the guarding mode. Contact guarding occurs at temperatures above 30° C (Corbet, 1999). In 2005, a ragged specimen, likely a wanderer from the mountains, was found at Sandpiper Pond in Eugene, Lane County, the first Willamette Valley record.

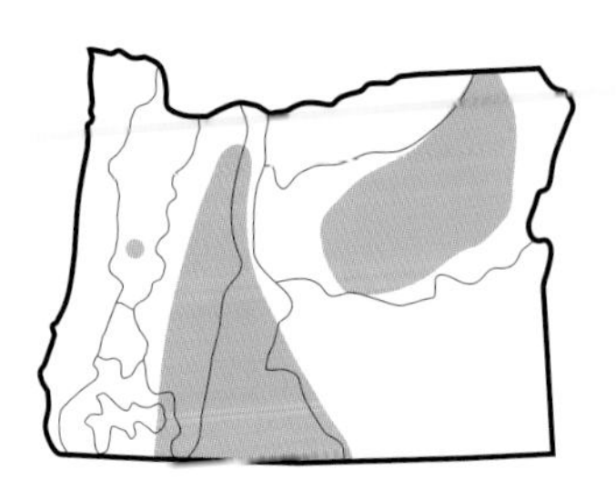

The adult flight season extends from mid-July to early October.

JAN	FEB	MAR	APR	MAY	JUN	JUL	AUG	SEP	OCT	NOV	DEC

White-faced Meadowhawk male

White-faced Meadowhawk female

Short: 30–39 mm

Striped Meadowhawk

Sympetrum pallipes

The Striped Meadowhawk shows two pale stripes on the sides of the brown thorax and small, pale stripes on the front of the thorax, and this species is named for these stripes. Look for these thoracic side stripes in the field to aid in identification. The red abdomen has small irregular black markings on the sides, but they are not triangular as on the White-faced, Band-winged, or Cherry-faced Meadowhawks. The eyes are dark reddish-brown on top and lighter on the bottom, the face is pale, and the legs are black. The pterostigma is reddish brown blending to lighter colors at the ends. The female has a light brown thorax with white side stripes. The eyes are dark brown on top and lighter on the bottom, the face is pale, and the abdomen is brown with black stripes along the upper sides. The legs are black. The sides of the abdomen may also show pruinosity on the front segments. The bases of the wings in both sexes are often bright red where they attach to the thorax.

The Striped Meadowhawk is a western species found from New Mexico into British Columbia and as far east as Nebraska. This common species frequents ponds and wooded streams throughout Oregon, even in urban areas. It can be very abundant at marshes, bogs, and seasonal wetlands in the mountains up to 7,500 feet in elevation.

Oviposition takes place in tandem and often in groups. It is not uncommon to see a dozen pairs ovipositing in an area a few feet in diameter, often onto grassy areas that are dry even when there are many less arid acres nearby.

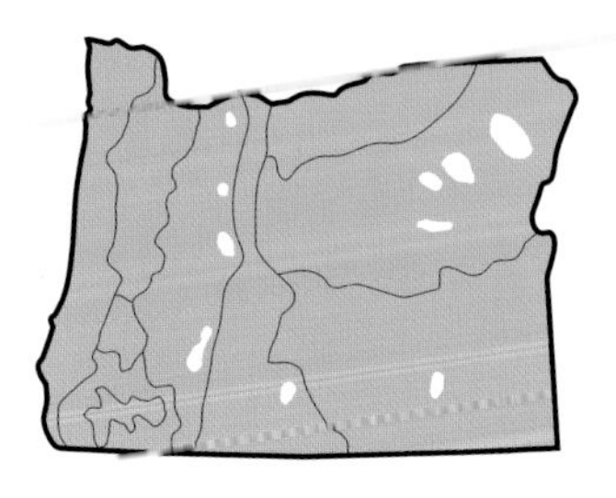

The adult flight season extends from early June to late November.

JAN	FEB	MAR	APR	MAY	JUN	JUL	AUG	SEP	OCT	NOV	DEC

Striped Meadowhawk male

Striped Meadowhawk female

Medium: 32–36 mm

Band-winged Meadowhawk

Sympetrum semicinctum

The basal half of the wing is amber, the source of its scientific and English names. The basal wing colors make this species easy to separate from the other meadowhawks in the field. The Band-winged Meadowhawk shows black lines on the sides of the reddish-brown to pale thorax. The front black line is jagged and shows as a flattened "W" pattern. The sides of the red abdomen show a series of long black triangles. The eyes are deep red on top and the legs are black. The female also has the basal half of the wing amber, with similar black markings on a thorax that is olive on top and brighter yellow low on the sides. The eyes are red on top and lighter below, and the face is pale. The abdomen is light olive with black markings similar to the male on the posterior segments. The male, female, and immature all have a thick black line on top of abdominal segments 8 and 9.

The Band-winged Meadowhawk is found across the U. S. and southern Canada. The western population was previously known as the Western Meadowhawk but has been combined with an eastern species (Band-winged Meadowhawk) as intermediates were found. The Band-winged Meadowhawk is common in marshy ponds and nearby wet meadows in eastern Oregon and in southern interior valleys as far north as the central Willamette Valley. It is found up to 7,500 feet in elevation. There are scattered records from Columbia, Clackamas, and Marion Counties. We have seen it very abundant in grassy areas at Fern Ridge near Eugene where it perches on grasses. It even forages into parks and gardens, especially in the southern Willamette Valley.

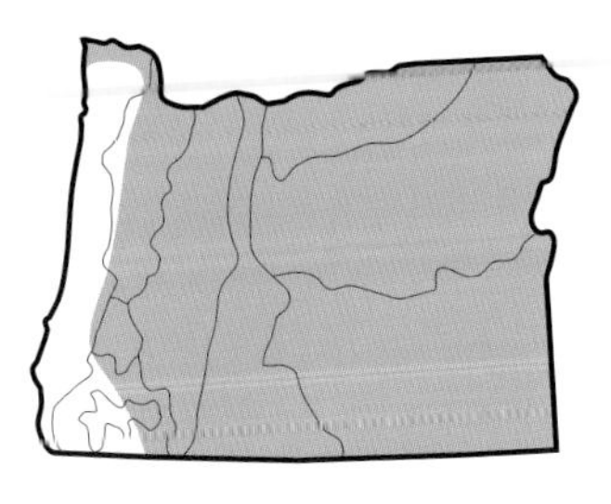

The adult flight season extends from late June to early October.

JAN	FEB	MAR	APR	MAY	JUN	JUL	AUG	SEP	OCT	NOV	DEC

Band-winged Meadowhawk male

Band-winged Meadowhawk female

Medium: 30–40 mm

Autumn Meadowhawk

Sympetrum vicinum

The Autumn Meadowhawk has an unmarked thorax and abdomen, which appear to be dull reddish-brown. The legs are reddish-brown, and the abdomen is extremely thin. In the field, the small size along with the very thin abdomen will aid in identification. The wings show a small amount of amber at the base, and the veins are uniformly dark. The similar Saffron-winged Meadowhawk is larger, has black low on the sides of the abdomen, and has black or black-streaked legs. Females are a uniform reddish-brown color including eyes, legs, and pterostigma. There is also an amber wash at the base of the wings and pruinosity on the underside of the abdomen. When perched, this female meadowhawk is easily identified by the large subgenital plate which projects down below abdominal segment 8. A cluster of eggs is sometimes visible in the space between the subgenital plate and the abdomen.

The Autumn Meadowhawk is found in the eastern and central United States and just into Canada. In the west, it occurs from central California into Canada, but is mostly absent from the Rocky Mountain States and southwest. It has, however, been found at a nearly 10,000-foot elevation in Colorado (Corbet, 1999). This uncommon to common meadowhawk is found near wooded streams, ponds, and lakes at scattered locations across Oregon. It is common below 3,000 feet in elevation and less common up to 5,000 feet. Look for it perched near well-vegetated ponds with nearby woods. This species was formerly named the Yellow-legged Meadowhawk but the common name was changed in 2004. The fact that its legs are not always yellow and that it is usually seen late in the flight season in September and October justify the name change.

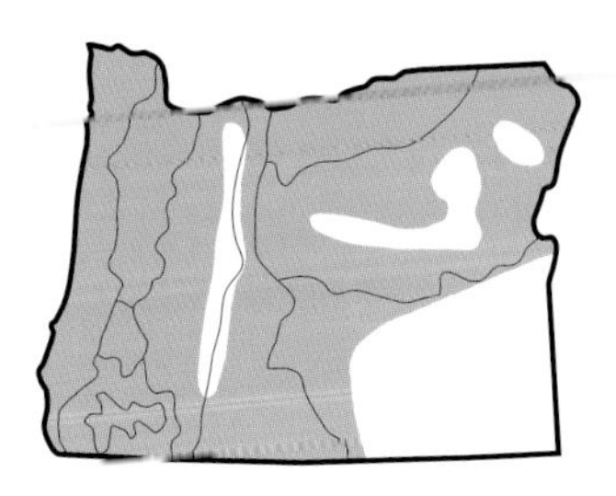

The adult flight season extends from early August to early December.

JAN	FEB	MAR	APR	MAY	JUN	JUL	AUG	SEP	OCT	NOV	DEC

Autumn Meadowhawk male

Autumn Meadowhawk female

Short: 31–37 mm

Wandering Glider

Pantala flavescens

The genus name, *Pantala*, literally means “all wing,” referring to the long, wide wings. These are also referenced by the term “glider.” The English name, “Wandering,” refers to the extensive migratory behavior of this species. The abdomen is a combination of yellow and orange— appearing golden overall (hence flavescens meaning yellowish)–with black medial markings along the top surrounded by lighter areas. The Wandering Glider has chestnut eyes which stand out even in flight. The juvenile’s face is golden-yellow, turning redder with maturity. When seen in flight, this combination of face, eye, and body color is noticeable. The Wandering Glider shows no dark wing markings, but may show an amber wash at the wingtip. The pterostigma is also a golden yellow. The female and male are very similar in color and pattern.

There are only four recordings of this visitor to Oregon including three from the Willamette Valley and a single record from Harney County. On August 13, 2005, we recorded six Wandering Gliders at Sandpiper Pond in Eugene—more than had been recorded in Oregon’s history. These included mating and ovipositing pairs; the first Oregon evidence of breeding. We observed mating pairs at Sandpiper Pond into September during 2005. Females oviposit in tandem with the male, making repeated dips to the water. It is unknown whether the Wandering Glider completes its life cycle in Oregon. The other known Willamette Valley sightings are from the Freeway Lakes Park on Oak Creek in Linn County. This broad-winged glider is known for its migrations and is adapted to exploiting seasonal habitat. Migrations are associated with storm fronts, and in southern China this species may be called “Typhoon Flies” for this reason (Corbet, 1999). We have seen it ovipositing on vehicles from the Mexican desert to a Midwestern parking lot, mistaking the shiny surface for water. It breeds on every continent except Europe and Antarctica (Needham, et. al., 2000).

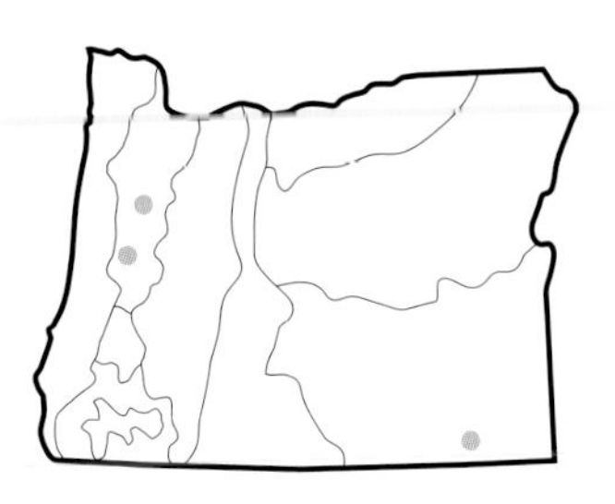

The adult flight season extends from early June to early September.

JAN FEB MAR APR MAY JUN JUL AUG SEP OCT NOV DEC

Wandering Glider male

Wandering Glider female

Medium: 46–50 mm

Spot-winged Glider
Pantala hymenaea

The Spot-winged Glider has a characteristic semicircular dark basal spot on the hindwing, the source of its English name. The wings are long and broad and well suited for migrating. The broad wings are the source of the scientific name *hymenaea* which means membrane. The body is patterned in browns and lighter spots. The face and eyes are a reddish-brown. The front of the thorax is olive brown, and the sides have olive brown surrounded by lighter areas. While the spot on the hindwing is the main feature helping distinguish the Spot-winged Glider from the Wandering Glider, the top of the abdomen of the Spot-winged Glider is more patterned—with a series of white spots bounded by black and then brown. When seen from the top, it gives less of an orange impression than the Wandering Glider. In flight, look for the red face and the wing spot. The female has brown eyes and a brown face. The front of the thorax is brown, the sides olive to brown, and the abdomen is darker and has a series of medial black spots surrounded by brown. The wing has the characteristic dark basal spot.

Oviposition takes place in tandem with the male, and seems to favor small pools over larger bodies of water. The female extrudes a ball of eggs as the pair repeatedly dips to tap the water. This glider has been known to complete the larvae stage in only 28 days (Corbet, 1999).

The Spot-winged Glider sporadically visits still waters in Oregon, migrating into the state from the south. We have one photographic record of this rare dragonfly from 2001, hanging up in a flower garden in the Friendly neighborhood in Eugene. In May, 2009, we found this glider ovipositing in pools in two locations near Medford. In addition, there were a number of males patrolling—indicating that a migrating flight had come into the area. While we don't have enough evidence to predict its irregular migrations, you should look for this species throughout Oregon.

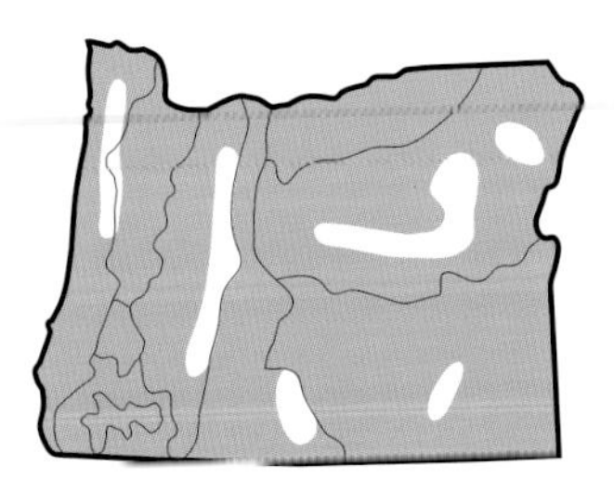

The adult flight season extends from mid-May to late August.

JAN	FEB	MAR	APR	MAY	JUN	JUL	AUG	SEP	OCT	NOV	DEC

Spot-winged Glider male

Spot-winged Glider female (inset: female top view)

Medium: 45–50 mm

Black Saddlebags

Tramea lacerata

The scientific name, *lacerata,* refers to the ragged or lacerated edges of the hindwing spots, and the English name, “saddlebags,” refers to its black hindwing spot. The male is truly a black dragonfly with clear wings, except for a large black “saddle” patch at the base of the broad hindwing. There are faint yellow spots on the top of the abdomen, often most visible on segment 7. These spots may not be visible on very mature males. The eyes are deep brown, and the face is metallic purple in mature males. The female is similarly marked, but is brown with yellow spots on the abdomen. These spots increase in size to segment 7. Spots may be scarcely visible on mature specimens. Black Saddlebags are strong fliers with long, broad wings, and they hang up when perched. You can easily learn to identify this common species in flight. With its large size and dark saddles, it casts quite a shadow as it flies overhead, and you may notice it first by the “Shadow of the saddlebags” on the ground.

Black Saddlebags have interesting ovipositing behavior. After mating, they fly in tandem over the water. The male will release the female and she will drop to the water’s surface to release eggs. Then she will fly back up and he will re-engage her and resume the tandem position, waiting to release her when she is ready to oviposit again. Watch for this fascinating behavior when you are in the field. This species frequents ponds and small lakes, but can be seen flying over fields, parks, and gardens. It is one of our common urban dragonflies, but has not been found in Tillamook and some eastern Oregon Counties. Black Saddlebags are known to migrate, and both migrant and resident populations occur in some areas (Corbet, 1999).

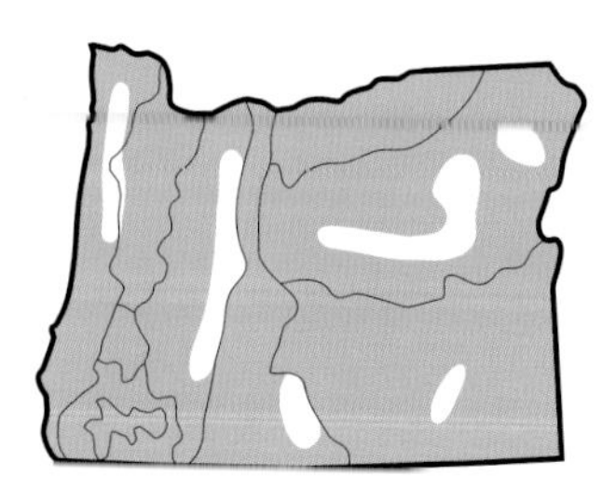

The adult flight season extends from mid-May to mid-October.

JAN	FEB	MAR	APR	MAY	JUN	JUL	AUG	SEP	OCT	NOV	DEC

Black Saddlebags male (inset: male front view)

Black Saddlebags female

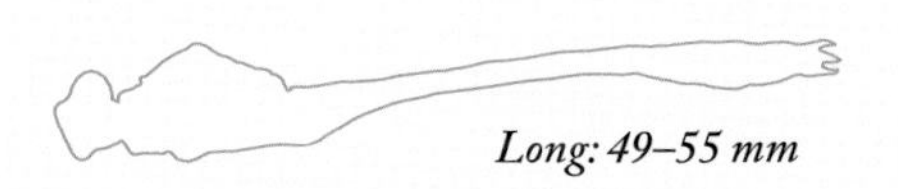

Long: 49–55 mm

Aztec Dancer copulating pair

Western Red Damsel male

B. Damselflies

Damselflies have eyes that are widely separated by more than one eye width with a "hammer-headed" appearance. They have a relatively thin thorax and abdomen. Their forewings and hindwings are shaped similarly. At rest, damselflies hold their wings folded above or next to the abdomen. However, in one family, the spreadwings, the wings are held horizontally and directed rearward, like a jet airplane's wings. Damselflies have two upper appendages (cerci) and two lower appendages (paraprocts). See the morphology section in the introduction for comparison of damselflies and dragonflies (Figure 2, page 22, and Identification Chart A).

The twenty-eight damselfly species are represented by three families in Oregon: broad-winged damsels—two species (Calopterygidae), spreadwings—seven species (Lestidae), and pond damsels—nineteen species (Coenagrionidae).

Broad-winged Damsels, Calopterygidae

The broad-winged damselflies are among the most attractive damsels found in North America. There are only eight species in this family in North America and all are stream dwelling. They include five species in the genus *Calopteryx* (jewelwings), and three species in the genus *Hetaerina* (rubyspots). Oregon has two species, one species in each of those two genera. These are among the most approachable and easily identified damselflies due to their striking color patterns. They have wings that are broader than other damselflies from the base to the tip, and the wings are held closed over the back when perching. The body colors are metallic green or red, and the wings have unmistakably colored sections. The colors often look different depending upon how they are situated in the sun.

Broad-winged damsel larva

The adults perch horizontally on long legs on vegetation, usually overhanging the water or quite near. They make slow flights out over the water and up and down the stream and are dramatic as the wings flash color. Male jewelwings act aggressively towards each other, at times appearing to be whirling around in tight circles up and down over the water. They have especially interesting mating and ovipositing behaviors. The broad-winged damselflies perform complex courtship displays, and males defend territories and display aggressive behavior towards intruders (Corbet, 1999). After mating, the female will crawl down the stems of plants and oviposit underwater by inserting the sharp ovipositor into plant stems. Oviposition usually takes place within a few minutes of mating (Corbet, 1999). Males may also be seen non-contact guarding females during oviposition, so if one sees an ovipositing female, look for a male perched above.

The long-legged larvae of this family cling to vegetation and roots along the margins of streams and rivers. They are quite dramatic in appearance and, with their streamlined shape, look well adapted to living in flowing water.

River Jewelwing

Calopteryx aequabilis

This spectacular species is metallic green on the thorax and abdomen and has very long legs. The head and thorax are very hairy, and the eyes are dark. The large wings have an iridescent black tip on the outer third. The genus name, *Calopteryx,* means beautiful wing. The female is a duller green and has a distinctive white pterostigma contrasting with the wing tip's black background. We have also seen females with the entire wing dark. Nevertheless, the white pterostigma of the female is characteristic. The superb colors of this species make it easy to identify in the field. Males defend territories and may display for females. The male may dive to the water surface to show the female an oviposition site, laying his wings flat on the surface of the water (Corbet, 1999).

The River Jewelwing is distributed across the northern United States and southern Canada. It is only known as far south as San Francisco on the west coast. It is found throughout Oregon up to 4,600 feet in elevation, but it has not yet been found in some coastal, north Willamette Valley, and eastern Oregon counties. The River Jewelwing perches on low vegetation or lower tree branches along small- and medium-sized rivers and streams—or, alternatively, on rocks amid the riffles. It prefers waters with abundant submerged vegetation and rootlets. The River Jewelwing is such a beautiful and appropriate name for this species. Along small, wooded streams, the sunlight filters through maples and alders and twinkles off the riffles, and these long damselflies flit through shafts of light on four long, black-tipped wings. Reflecting sparkles of metallic green, the jewelwings seem like fairies dancing along the stream course. Mount Pisgah Arboretum at Buford Park near Eugene is a good spot to see this species along the Coast Fork of the Willamette River.

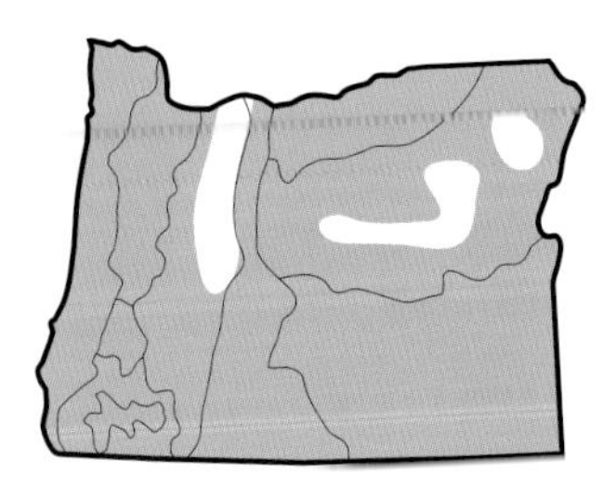

Adults have been found flying in Oregon from mid-May through mid-September, but are more common earlier in the season.

JAN	FEB	MAR	APR	MAY	JUN	JUL	AUG	SEP	OCT	NOV	DEC

River Jewelwing male

River Jewelwing female

Long: 45–50 mm

American Rubyspot

Hetaerina americana

The thorax of the male of this attractive damsel is reddish-black and the abdomen is metallic brown. The body has an iridescent reddish hue. The eyes and head are dark and somewhat hairy, as is the thorax. The wings are ruby-red at the base extending to just short of the nodus and are clear beyond. The red in the wing is brighter on the top side than the bottom, so is more obvious when in flight. When the red wing spot flashes in flight, the American Rubyspot is unmistakable, and unmistakably beautiful. This flash of color is also likely important in defending a territory, as males often drive mated pairs from a territory (Corbet, 1999). The female has a metallic reddish-brown thorax with several cream-colored stripes on the sides. The abdomen is also metallic reddish-brown above and a cream color below. The wings have a reddish-brown hue at the base but not the bright red color of the male. The wing beyond the nodus may have an opaque brown tint. The female is also quite an attention-getting insect.

The American Rubyspot is found across the U. S., but is not known from Idaho or Washington. In Oregon, it has a limited distribution and has only been found in Marion, Benton, Lane, Linn, Douglas (all below 1,000 feet in elevation), and Klamath (at 3,150 feet) Counties. It is more common in the southern Willamette Valley than further north. In Lane County it can be found along the Coast Fork Willamette River at Mt. Pisgah, and this is a good place to see it (See also the South Umpqua River Dragonflying Spot for another good location to find this species.) Look for it perched on willows overhanging the stream. At Mt. Pisgah, it is very common along a small side channel of the river just past the park buildings. It appears to prefer this habitat over the larger river. There are historic records from Benton, Linn, and Marion Counties from 1936. On June 15, 2005, we found a female on willows along the Willamette River in Bryant Park in Albany, Linn County, near the mouth of the Calapooya River but have not found it on return visits.

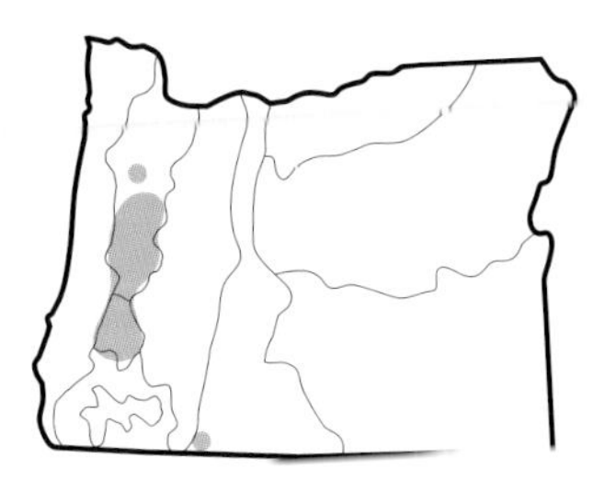

The flight period and habitat are similar to the River Jewelwing, with the flight season extending from early May to early October.

JAN	FEB	MAR	APR	MAY	JUN	JUL	AUG	SEP	OCT	NOV	DEC

American Rubyspot male

American Rubyspot female

Long: 44–45 mm

Spreadwings, Family Lestidae

Oregon has seven of the 19 North American spreadwings in two genera, *Archilestes* and *Lestes* (see Identification Chart O). *Archilestes* are known as the stream spreadwings and the *Lestes* are known as the pond spreadwings. This distinction is not always applicable: sometimes the California Spreadwing inhabits ponds, and many of the pond spreadwings can be found in vegetation along streams.

These are rather large damselflies and, unlike our other damselflies, they tend to hang up and hold their wings open, more like dragonflies. However, spreadwings hold their wings slightly above horizontal and angled back toward the abdomen, like a jet's wings. At night, in cold weather, or when threatened, all spreadwings are known to close their wings (Paulson, 2009). They have long pterostigma; longer than their eye width. Spreadwings also have very long leg spines.

Spreadwing larva

The females saw into plants to lay their eggs just above or below the water surface. Females oviposit in tandem with the males. Studies of development of spreadwing eggs and larvae of spreadwings illustrate how they successfully exploit many habitats. Eggs may have a diapause to allow them to survive unfavorable periods, and larvae are capable of surviving after being frozen in ice (Corbet, 1999). The larvae are long and thin with long legs. The larvae swim in open water like minnows. Spreadwings can be found at many temporary and permanent waters in Oregon. The male terminal appendages are useful for determining species.

Figure 6 • Northern vs. Sweetflag Spreadwings: Telling Them Apart

Key Features	Northern (*Lestes disjunctus*)	Sweetflag (*Lestes forcipatus*)
Abdominal Segment 2	Entirely pruinose	Basal 2/3 pruinose
Upper Appendages (Cerci) (top view)	Distal tooth; Basal tooth. Basal tooth farther from S10; distance between teeth short; distal tooth large, sharp	Basal tooth nearer to S10; distance between teeth long; distal tooth small, often blunt

Identification Chart O • Spreadwings, Genera: *Archilestes* and *Lestes*

Species (size)	♂ Appendages (bottom view)	Notes	♀ Ovipositor (side view)
California *Archilestes californica* Very Long 45–50 mm	Short, stubby	♂ and ♀ very large—our largest spreadwing; pterostigma light brown	Bulbous abdomen tip without pruinosity
Spotted *Lestes congener* Long 36–42 mm	Short	♂ and ♀ have 2–4 dark, elongated spots low on thorax (bottom view)	
Northern *L. disjunctus* Long 35–40 mm	Long, sometimes crossed	♂ and ♀ no spots under thorax; S2 completely pruinose; (See Figure 6)	
Sweetflag *L. forcipatus* Long 34–40	Long	Heavily pruinose overall; basal 2/3 of S2 pruinose; (See Figure 6)	Extends beyond terminal tip
Emerald *L. dryas* Long 32–37 mm	Medium, expanded tips pointed inward	Bright, metallic green atop thorax and abdomen	Reaches terminal tip
Black *L. stultus* Long 35–37 mm	Medium, expanded tips pointed inward	Black atop abdomen (our darkest spreadwing)	
Lyre-tipped *L. unguiculatus* Long 35–37 mm	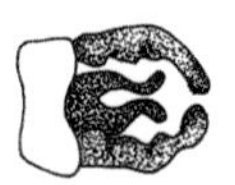 Lyre-shaped, "unique"	Back of head pale (others darker); pterostigma very pale, white to light brown	

California Spreadwing

Archilestes californica

This is our largest damselfly and can be identified by its size alone. The sides of the thorax have a white side stripe bordered by two black stripes. The abdomen is tan with thin black bands and a contrasting pruinose bluish-gray tip. The eyes are blue, and the pterostigma is white to light brown. The female lacks the bright blue eyes of the male and the pruinose abdominal tip, but otherwise the female and male are similar including the light-colored pterostigma.

The California Spreadwing is a western species and is found from Mexico to northern Washington. It can be found throughout Oregon up to 5,400 feet in elevation and is fairly common in the Willamette Valley. It prefers small, slow-moving streams bordered by willows, but can range into yards and gardens. This damselfly can also be common at ponds, like Grimes and Sandpiper Ponds in Eugene, which have willows and cottonwoods along the shoreline. Tandem pairs are skittish and often fly up when approached and will land at eye level or higher in vegetation. It is impressive to see a tandem pair flying up into the vegetation along a stream or pond; they are unmistakable due to their large size.

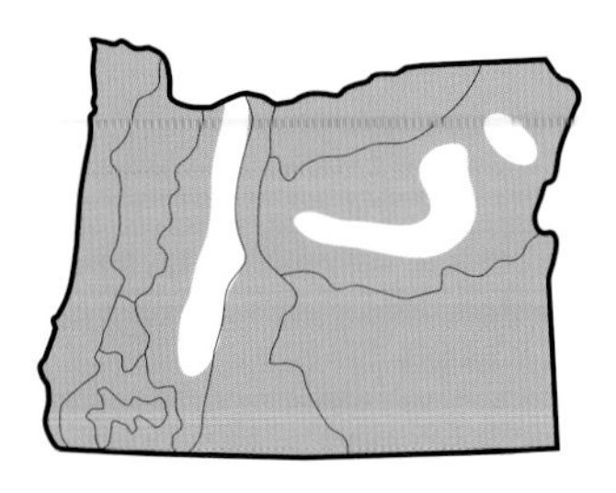

In Oregon, the California Spreadwing's adult flight season is rather late. It is common in August and September, although it has been found from mid-June through mid-November.

JAN	FEB	MAR	APR	MAY	JUN	JUL	AUG	SEP	OCT	NOV	DEC

California Spreadwing male

California Spreadwing female

Very Long: 44–50 mm

Spotted Spreadwing

Lestes congener

The thorax of the Spotted Spreadwing is black on top, with narrow, light-brown shoulder stripes and wide white bands at the rear of the thorax. There are two to four dark spots on the underside of the white thorax (see photographic inset), which account for the common name. The eyes are blue and the abdomen has a bluish pruinose top. When viewed from the top, the Spotted Spreadwing's lower abdominal appendages are short. In comparison, the similar Northern Spreadwing has relatively longer lower appendages. The female's thoracic pattern and color is similar, including the four dark spots under the thorax, but the female lacks the blue eyes and the pruinose tip on the abdomen.

The Spotted Spreadwing is found all across North America as far north as the Northwest Territories of Canada. It is common throughout Oregon up to 5,700 feet in elevation and can be seen throughout the Willamette Valley at well-vegetated ponds, lakes, and slow-moving streams. In early September, we have encountered hundreds of Spotted Spreadwings along the Coyote Creek Nature Trail south of Fern Ridge in Lane County; they almost outnumber the mosquitoes. This damselfly's habitat also includes stagnant, seasonal, and saline waters (Westfall and May, 2006).

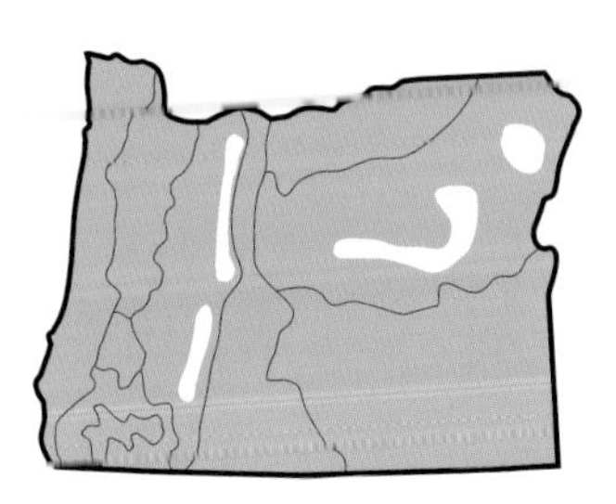

The adult flight season extends from late May to mid-November.

JAN	FEB	MAR	APR	MAY	JUN	JUL	AUG	SEP	OCT	NOV	DEC

Spotted Spreadwing male (inset: spots on underside of thorax)

Spotted Spreadwing female

Long: 36–42 mm

Northern Spreadwing

Lestes disjunctus

The male Northern Spreadwing's thorax is pruinose blue with blue stripes which may be dark enough to appear black. The first two and last three abdominal segments are pruinose blue. The eyes are blue, and the back of the head is black. The male's lower abdominal appendages are long and straight when viewed from the top. Note that the lower abdominal appendages can sometimes be crossed. Without a hand lens, these crossed appendages could be confused with those of the Lyre-tipped Spreadwing. The female has blue eyes and a dark thorax with a narrow pale stripe on the shoulder of the thorax and a partial stripe on the side. The base and tip of the female abdomen are also pruinose. The amount of pruinosity in both sexes depends upon the maturity of the individual.

The Northern and Sweetflag Spreadwings are very similar species. Check the amount of pruinosity on the top of abdominal segment 2 and the teeth on the inside of the upper appendages (cerci) to tell these species apart (see Figure 6, page 226).

The Northern Spreadwing is found from the southwestern United States as far north as Alaska and northern Canada. It is found across Oregon up to 7,500 feet in elevation and can be seen in emergent vegetation near lakes, ponds, and streams.

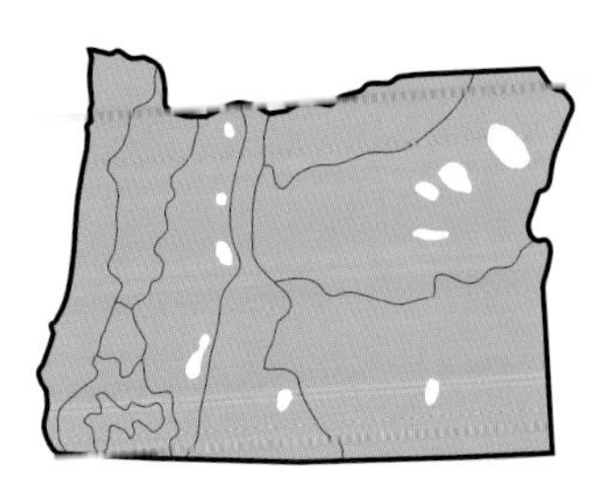

This species was formerly named the Common Spreadwing.

The adult flight season extends from late May to late September.

JAN FEB MAR APR MAY JUN JUL AUG SEP OCT NOV DEC

Northern Spreadwing male

Northern Spreadwing female

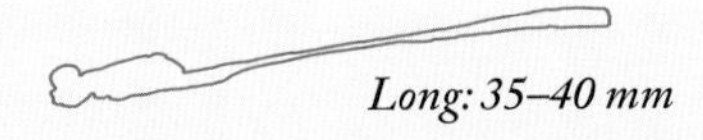

Long: 35–40 mm

Emerald Spreadwing

Lestes dryas

The top of the thorax and abdomen of the Emerald Spreadwing are bright, metallic green, making this species easy to identify in the field. The side of the thorax is also green and the eyes are blue. The first two and last two abdominal segments are pruinose blue with the intervening segments green on top and cream colored below. The male's appendages are long and they broaden at the tip. The female may be metallic green or copper colored on both the thorax and abdomen. The shoulder of the thorax has a narrow cream colored stripe, and the lower portion of the thorax and abdomen are also a cream color. The eyes are gray and the top of the head is dark. The metallic colors of both sexes of Emerald Spreadwing are striking when one encounters them in the field.

This is another species that has a wide distribution across North America from Alaska to the southwestern United States. The Emerald Spreadwing is common at higher-elevation lakes, marshes, seasonal wetlands, and meadows in Oregon. It has been found in the southern Willamette Valley and is common at some locations, but is largely absent from the northern Valley.

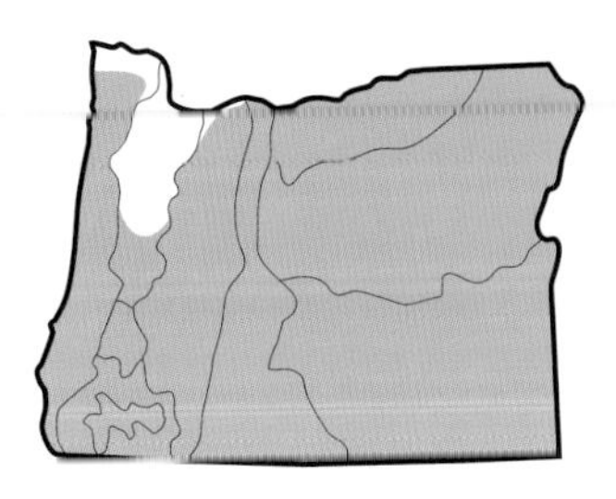

The adult flight season extends from mid-May to early November.

JAN FEB MAR APR MAY JUN JUL AUG SEP OCT NOV DEC

Emerald Spreadwing male

Emerald Spreadwing androchrome female (inset: gynochrome female)

Long: 32–37 mm

Sweetflag Spreadwing
Lestes forcipatus

The Sweetflag Spreadwing male is more pruinose when mature than any of our other spreadwings. The thorax is entirely blue-gray and pruinose along with abdominal segments 1, a portion of segment 2, and segments 8, 9, and 10. The top and sides of the remainder of the abdominal segments are black with some gray pruinosity tapering along the sides. The eyes are turquoise and the legs are black. The male has terminal abdominal appendages identical to the Northern Spreadwing. However, they can be separated in the field by noting the pruinosity on the top of second abdominal segment (see Figure 6, page 226). The Sweetflag Spreadwing is pruinose on the forward two-thirds of the segment while the Northern Spreadwing's second segment is entirely pruinose. The female Sweetflag Spreadwing has thin, dark stripes along the center and shoulders of the thorax, separated by grayish colors becoming pruinose in maturity. The abdomen is blue-gray on the sides and black on the top. The ovipositor is very large and impressive compared to our other spreadwings and extends beyond the end of the abdomen. This can be observed even in copulating pairs, so observation of pairs can assist in learning to identify this spreadwing in the field.

The Sweetflag Spreadwing was first found in Oregon in 2009 and is currently known only from Union and Wallowa Counties. It will undoubtedly be found in more locations in the future. It occurs from Nova Scotia to Virginia in the east and ranges to the west coast. Its distribution is spotty in the west due to confusion in identification with the Northern Spreadwing (Paulson, 2009). In Oregon, it has been found in heavily vegetated ponds and wetlands. Mating pairs can be observed ovipositing in stems of rushes and sedges from a few feet above water to the water surface. In eastern North America, the female lays eggs in the aquatic Sweetflag, an iris, thus its name.

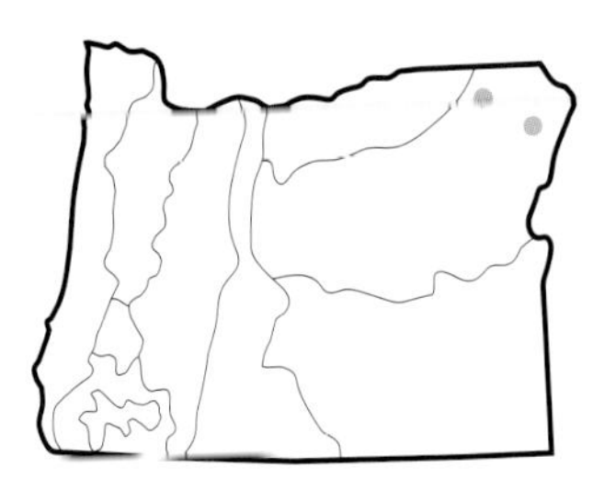

In Washington, the adult flight season extends from July into September (Paulson, 2009), and its Oregon adult flight season will expand as more observations are recorded.

JAN	FEB	MAR	APR	MAY	JUN	JUL	AUG	SEP	OCT	NOV	DEC

Sweetflag Spreadwing male

Sweetflag Spreadwing female

Long: 34–40 mm

Black Spreadwing

Lestes stultus

The Black Spreadwing male has an overall dark metallic color. It is the darkest of the Oregon spreadwings. The eyes are blue, the top of the head is black, and the thorax has a coppery sheen with very thin shoulder stripe. The side of the thorax is pale colored at the rear. The abdomen is dark metallic above and may be metallic green with a blue pruinose tip. The pterostigma is black with a thin white line on the sides. The male abdominal appendages are identical to the Emerald Spreadwing. Females and males are colored similarly, but the female has brown eyes and more extensive pale areas on the lower portion of the abdomen.

Johnson (2006) discusses questions related to the status of the Black Spreadwing which is structurally identical to the Emerald Spreadwing. The only difference between these species is in color of the thorax, flight season, and elevation at which they are found. He also describes a site in Oregon where color blends between the species were found. For these reasons, Johnson indicates that the status of the Black Spreadwing needs further study to determine whether it is indeed a separate species or merely a color variation of the Emerald Spreadwing. We include it here as it is listed as a separate species by DSA.

The Black Spreadwing is only known from California and from Douglas, Jackson, and Josephine Counties in Oregon at low elevation valley locations. Manolis (2003) indicates that the habitat is temporary ponds and marshes in California, and this is the habitat where it is found in Oregon.

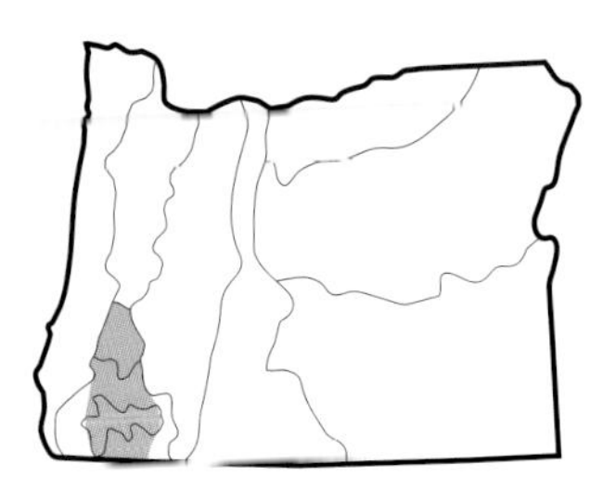

In Oregon, the Black Spreadwing has been found from mid-June to mid-July.

JAN FEB MAR APR MAY JUN JUL AUG SEP OCT NOV DEC

Black Spreadwing male

Black Spreadwing female

Long: 35–37 mm

Lyre-tipped Spreadwing

Lestes unguiculatus

This spreadwing has a coppery-bronze and metallic green color on its upper thorax and a metallic green sheen on its upper abdomen. The middle abdominal segments are pale blue on the bottom. These colors may appear more bronze or green depending on the incident light. The eyes are blue, and the thorax has a pale green shoulder stripe. The thorax is light blue on the posterior section, and the abdomen has a pruinose blue tip. Its pterostigma has a dark center with pale, creamy outer edges. When viewed from the top, its terminal appendages are distinctive—with the lower appendages curving inward and then outward to give it a "lyre-shaped" appearance, and thus its name. The female has similar colors but the eyes are dark, the thoracic side stripe is broader and pale green, and the abdomen lacks extensive pruinosity but the last three segments may be slightly pruinose. The lower half of the abdomen is also a pale cream or pale green.

The Lyre-tipped Spreadwing is found across central North America and is common across much of Oregon up to 7,500 feet in elevation, but it has not been recorded along the coast or in Wallowa County. It prefers emergent vegetation on the edges of ponds, lakes, and slow-moving streams.

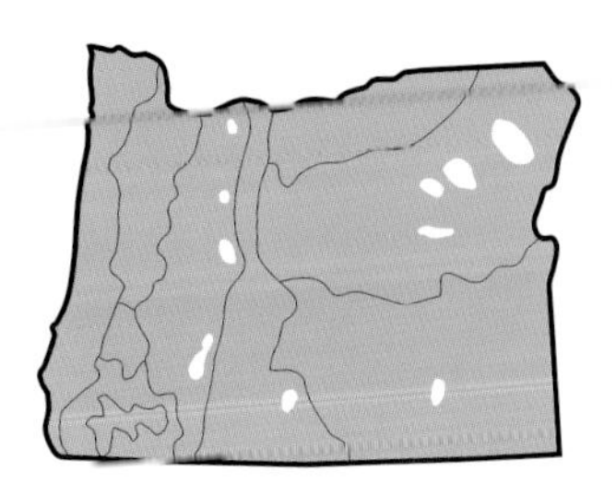

The long adult flight season extends from mid-May to late September.

JAN	FEB	MAR	APR	MAY	JUN	JUL	AUG	SEP	OCT	NOV	DEC

Lyre-tipped Spreadwing male

Lyre-tipped Spreadwing female

Long: 35–37 mm

Pond Damsels, Family Coenagrionidae

This is the largest North American (103 species) and Oregon (19 species) family of damselflies, and its members are distributed worldwide. About two-thirds (68%) of Oregon's twenty-eight damselfly species are members of this family. Oregon's species fall into six different genera: red damsels, *Amphiagrion* (1 species); dancers, *Argia* (6 species); Eurasian bluets, *Coenagrion* (1 species); American bluets, *Enallagma* (7 species); forktails, *Ischnura* (4 species); and sprites, *Nehalennia* (1 species).

Pond Damsel (dancer) larva

The short to medium-sized pond damsels have narrow wing bases, but other families do also, so this is not a good defining feature. Most show spots behind the eyes and a bar at the back of the head between the eye spots. Unlike the spreadwings, pond damsels close their wings over the abdomen when at rest. While the Western Red Damsel is unique among Oregon damselflies with its red coloration, the others are black and blue or green. The generally black and tan females often have two color forms: 1) like the males (androchromatic), and 2) unlike the males (gynochromatic). The pond damselflies often oviposit in tandem in plant tissue and females have an ovipositor. Pond damselflies frequent ponds and marshes, but many species are associated with streams.

Dancers, Genus *Argia*

The dancers generally have strikingly identifiable colors in deep blues and lavenders (see Identification Chart P). However, we have one large dancer, the Sooty Dancer, that is a slate gray or a dark blue color. Some of our blue dancers and bluets could be confused in field observations, as the colors can be similar. In the hand, the tiny spines on the dancers' legs are longer than the space between spines; on bluets, the spine length is about the same as the space between spines (see Figure 7, page 245). This is a diverse genus with 32 species found in North America, and we have six of those species in Oregon. Two of our dancer species, the California and Aztec, are difficult to differentiate in the field. However, the Aztec Dancer has a very restricted range in the state. When encountering dancers in the field, one way to help identify them is to notice whether the top of the abdomen is mostly black: Paiute and Sooty; lavender: Emma's; or mostly blue: California, Aztec, and Vivid. The dancer identification chart is organized in that fashion.

Dancers are named for their bouncy flight style. When at rest, dancers' wings are usually folded and held above the abdomen—rather than lower over or along the side of the abdomen like the bluets. In addition, notice when dancers return to perch, they often spread their wings open and then snap them shut. Dancers often perch on the ground, rocks, or logs, rather than in vegetation like bluets, and can often be seen perching in sunny spots on trails near streams. They are associated with flowing water—from small streams to rivers—and generally oviposit in tandem. As with many other damselfly genera, female dancers have multiple color forms.

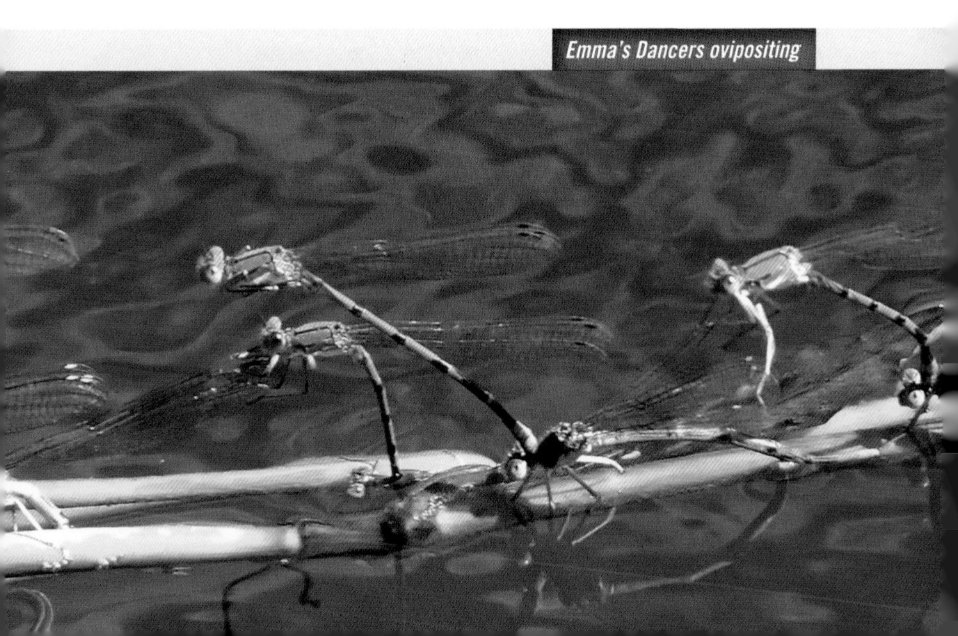
Emma's Dancers ovipositing

Identification Chart P • Dancers, Genus *Argia*

Species (size)	Thorax Stripes (side view)	Abdomen (top view)	Segment 10 (side view)
Dark atop Thorax and Abdomen			
Paiute *Argia alberta* Very Short 27–32 mm	Wide, forked stripe	Mostly black; S8–10 blue	Blue above, black below
Sooty *A. lugens* Very Long 40–50 mm	 Very dark, sooty	 Mostly black; S8–10 dark	 All black
Lavender on Thorax and Abdomen:			
Emma's *A. emma* Medium 35–38 mm	 Lavender; side stripe connected by thin line	 Lavender; S8–10 blue	 Blue above; light gray below with black marks
Blue atop Thorax and Abdomen:			
California *A. agrioides* Short 30–34	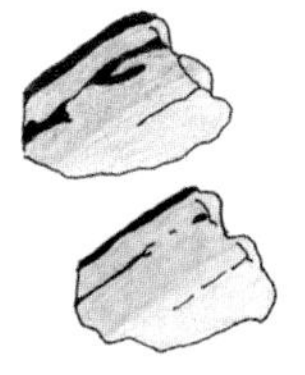	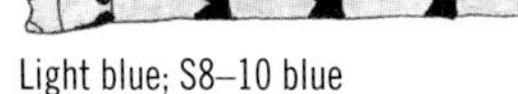 Light blue; S8–10 blue	
Aztec *A. nahuana* Short 30–35 mm	Two variations: 1) side stripe forked 2) side stripe faint, broken	Species similar (see Figure 8)	Blue above, gray below; very little black marking (See Figure 8)
Vivid *A. vivida* Medium 34–38 mm	 Deep blue; two broken side stripes	Deep blue; S8–10 light blue (top view) Note black arrows (side view)	 All blue

Figure 7 • Dancers (*Argia*) and American Bluets (*Enallagma*) Contrasts

Dancers, *Argia*	**Bluets, *Enallagma***
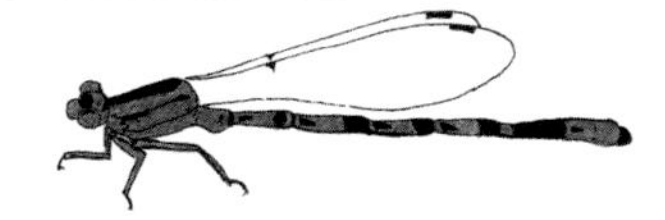 When perched, wings folded above the abdomen	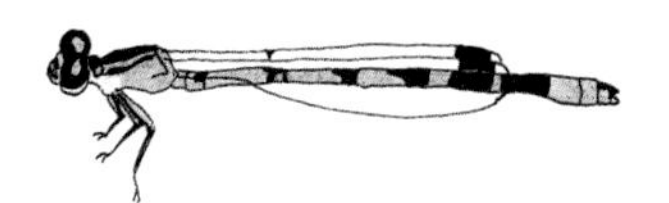 When perched, wings folded alongside the abdomen
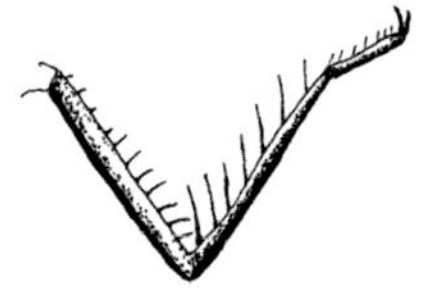 Tibial spines are longer than the spaces between spines	 Tibial spines are spaced about as far apart as the length of the spines
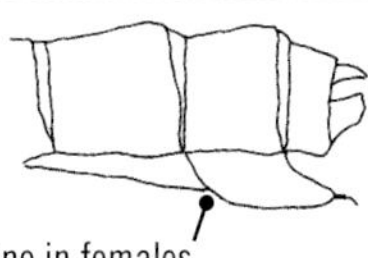 No vulvar spine in females	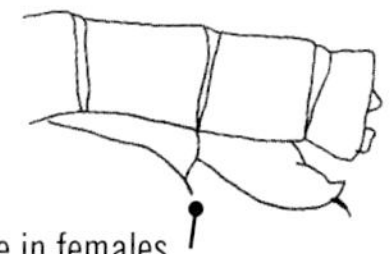 Vulvar spine in females (Note: no vulvar spine in female Taiga Bluet)

Figure 8 • California vs. Aztec Dancer ♂ Appendages

California Dancer *Argia agriodes*	**Aztec Dancer** *Argia nahuana*
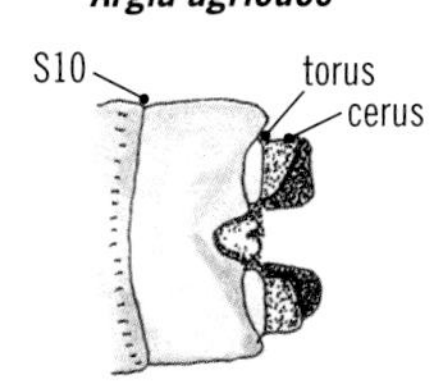 Tori divergent, about one torus width apart (top view)	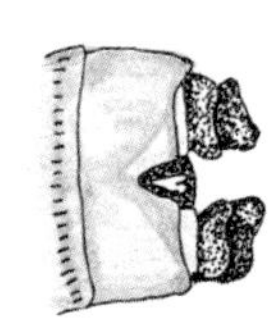 Tori parallel, less than one torus width apart (top view)
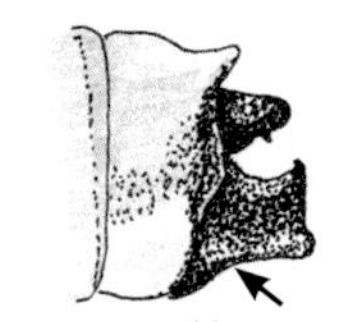 Lower appendage thinner overall (side view)	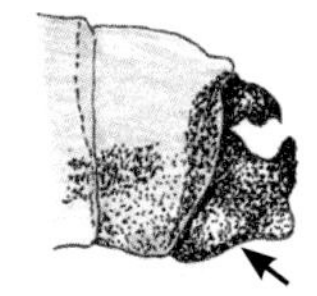 Lower appendage more robust (side view)

California Dancer

Argia agrioides

The male is bright blue but is not as rich or dark a blue as the Vivid Dancer. The California and Aztec Dancers have identical field marks and can only be differentiated in the hand by comparing the terminal abdominal appendages (see Figure 8, page 245). The head and face are blue as is most of the thorax. The thorax has a mid-dorsal black stripe. The sides of the thorax have a black stripe that is Y-shaped with the upper portion elongate. The lower branch of the Y is shorter and may be much reduced in some specimens. Abdominal segments 1 and 2 are blue with an upturned black spot on the side of segment 2. The middle abdominal segments are blue with black rings at the rear, and abdominal segments 8–10 are blue. The female is brown to tan in color. The thorax has black markings similar to the male, and the Y-shaped stripe may be greatly reduced. The female's abdomen has a black stripe on top which may be solid or interrupted, and has black markings on the sides towards the rear of the abdominal segments.

The California Dancer is found through southern Oregon as far north as Crook County. Oregon represents the northernmost extension of its range. It is found in Nevada, Arizona, Mexico, and California, as its name implies. Its behavior is similar to other dancers. Look for it perched in the open on rocks, the ground, or vegetation in the Rogue and Umpqua River valleys below 1,200 feet in elevation and in Malheur County from 3,200–4,000 feet.

The adult flight season extends from mid-June to late September.

JAN	FEB	MAR	APR	MAY	JUN	JUL	AUG	SEP	OCT	NOV	DEC

California Dancer male

California Dancer female

Short: 30–34 mm

Paiute Dancer

Argia alberta

The Paiute Dancer is very small (about an inch long) and has blue and black markings. The thorax has a black dorsal stripe bordered by blue. The shoulders of the thorax have a thick, black, Y-shaped stripe. This stripe is thicker than on the California Dancer. The top of the abdominal segments are mostly black with some blue lower on the sides of the abdomen. Abdominal segments 8–10 are blue on the top. The width of the stripe on the shoulders of the thorax and the extensive black on the abdomen make the Paiute Dancer easy to separate from the California Dancer in the field. Females are brown as are teneral males. The thoracic patterns of the female and male are similar, but the top of the female's abdomen is black, and it may have more brown depending upon maturity. Light blue females are also known (Westfall and May, 2006), but we have not observed them.

The Paiute Dancer is a species of the Great Basin and southwest and is named for the Paiute Tribe of that region. In Oregon, it is only found in Lake, Harney, and Malheur Counties up to approximately 4,600 feet in elevation where it is associated with hot springs. They can be found perched on the ground or in low, dense vegetation around the hot springs.

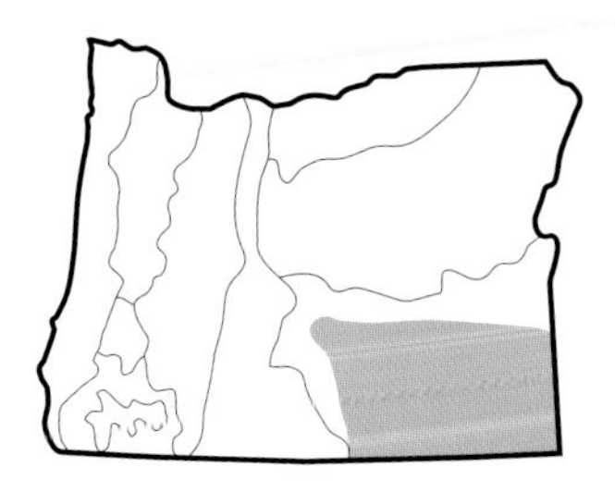

The adult flight season extends from late April to mid-September.

JAN FEB MAR APR MAY JUN JUL AUG SEP OCT NOV DEC

Paiute Dancer male

Paiute Dancer female

Short: 27–32 mm

Emma's Dancer

Argia emma

The Emma's Dancer male is a beautiful lavender color. It has a narrow black stripe on top of the thorax and an indented, narrow black stripe on the side. The abdomen is lavender with black rings and is blue on segments 8 and 9 and the top of 10. Abdominal segments 6 and 7 are mostly black.

In the female, brown replaces the lavender color of the male. The female and male have a similar thoracic pattern, and the female abdomen can have an olive tint on the middle segments.

Emma's Dancer is a western species found as far east as Iowa. The type locale for this species was from Satus Creek in Yakima County, Washington in 1915. It is common and local in Oregon at rocky rivers and larger streams that are sparsely vegetated, up to 5,800 feet in elevation. In Lane County it can be found at Mt. Pisgah near the Coast Fork Willamette River, where it is very common perching in sunlight along the trails. Here we have also observed a couple dozen pairs in tandem in submerged vegetation on a still backwater. Two dozen males hovered while hanging on to females as they lowered them into the Coast Fork waters and down to the plants where the eggs were laid. We have observed the same phenomenon along the John Day River in Wheeler County, but there over one hundred pairs were involved in similar behaviors.

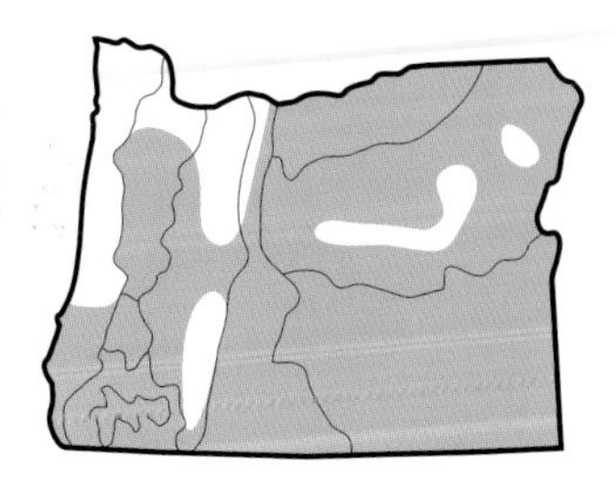

The adult flight season extends from late May to late September.

JAN FEB MAR APR MAY JUN JUL AUG SEP OCT NOV DEC

Emma's Dancer male

Emma's Dancer female

Medium: 35–38 mm

Sooty Dancer

Argia lugens

The Sooty Dancer is our largest dancer, and is appropriately named after its dark coloration. The male is a dark gray color and pruinose giving it a sooty appearance. Tim Manolis describes it aptly as having "... slate gray or blue gray pruinescence, appearing as if dipped in soot" (Manolis, 2003). There may be tinges of blue on the edges of the thorax. The entire body, including the abdominal tip, shows the dark color but the middle abdominal segments have small brown rings. These rings may be more or less visible depending on the maturity of the individual. The eyes appear black. Females and immature males have a brown thorax with black striping. We have observed mature females that were also blue with a black mid-dorsal line and a black Y-shaped stripe on the shoulder of the thorax. The abdomen is dark above and bluish below. At some sites, the blue females were the most common female color form. In addition, females colored much like the mature, sooty dark male, described above, have been observed.

The Sooty Dancer is found throughout the southwest and California. In Oregon, it is only known from Lake and Josephine Counties. Along the Illinois River in Josephine County, it is found from 1,100 to 1,400 feet in elevation, and along Twentymile Creek in Lake County it occurs at about 4,500 feet in elevation. It is common and easy to see at some sites in Josephine County. Look for the Sooty Dancer in open rocky areas and gravel bars with plenty of sun. They especially favor perch sites on emergent rocks in the stream bed.

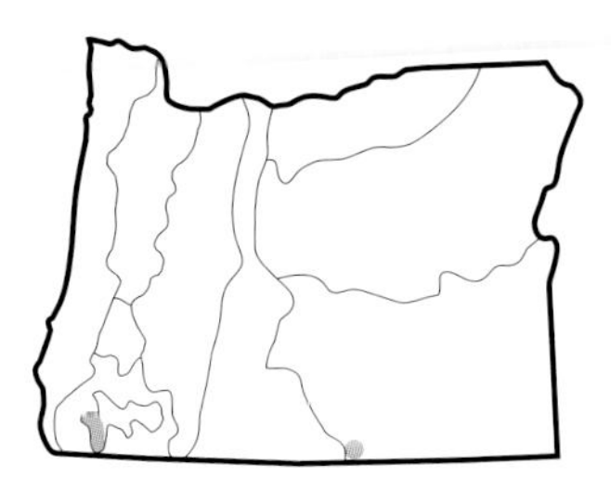

The adult flight season extends from mid-June to mid-October.

JAN FEB MAR APR MAY JUN JUL AUG SEP OCT NOV DEC

Sooty Dancer male

Sooty Dancer female (inset: androchrome female)

Aztec Dancer

Argia nahuana

Both the scientific and common names of this dancer relate to the Nahuatl group of native people from the Mexico City region, the type location. Review the description of the California Dancer as the color and patterns are virtually identical between the two species. The Aztec Dancer is very rare in Oregon and cannot be separated in the field from the more common California Dancer without close examination in the hand. The differences in the male genitalia are subtle. A hand lens is helpful but microscopic examination may be required. Refer to Figure 8, page 245, for differences between the two species' terminal appendages. These two species are so similar that the Aztec Dancer was considered a subspecies of the California Dancer at one time. When comparing the California and Aztec Dancer males in hand, examine the appendages in top view. The tori (singular torus) are thick pads that lie between the top of segment 10 and the upper appendages (cerci). In the California Dancer, the tori are angled outward and the tips are separated by about the width of a torus. In the Aztec Dancer, the tori are parallel and less than one torus width apart.

The Aztec Dancer is found in the arid regions of the southwest from Mexico north into California. It has a very localized distribution in Oregon, having been found by Jim Johnson and Steve Valley only at a hot spring along Twentymile Creek south of Adel in Lake County. This is the northernmost site within its known range. It should be considered a possibility at hot springs near the California border. We list this remote location (see the Dragonflying Spots section) where the Aztec Dancer was found, but one might be advised to look for it in California where it is more common.

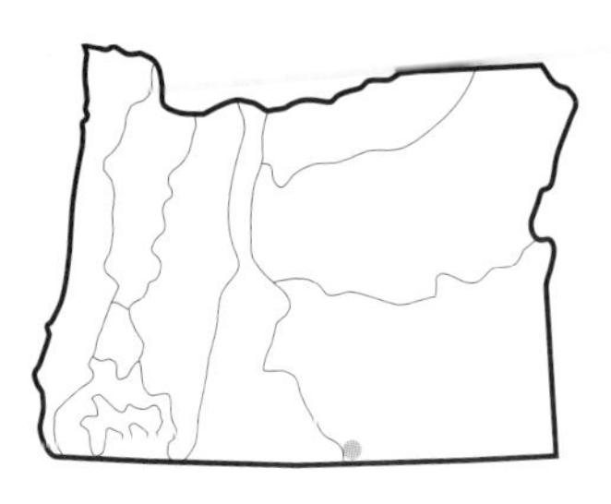

The adult flight season extends from early August to late September.

JAN	FEB	MAR	APR	MAY	JUN	JUL	AUG	SEP	OCT	NOV	DEC
							▬	▬			

Aztec Dancer male

Aztec Dancer female

Short: 30–35 mm

Vivid Dancer

Argia vivida

The Vivid Dancer's larger size, along with its behavior, helps distinguish it from our bluets and forktails. Males are colored vividly in bright, deep blue or violet-blue with black markings, thus giving this species its scientific and English names. The thorax has a broad, black stripe on top and a narrow black stripe on the side that is very narrow in the middle. The abdomen has broad blue bands with narrow black bands and arrow-shaped spots on the side with the arrow pointed toward the abdominal tip. Abdominal segments 8, 9, and 10 are blue. Females and males are patterned similarly, and the female may be brown, lavender, white, or blue depending on maturity and color form. Brown females have olive shading on the middle abdominal segments. Blue females are not as bright as blue males, have pale colored middle abdominal segments, and have abdominal segments 8, 9 and 10 blue. Look for the arrowhead markings on the abdomen to aid in distinguishing the Vivid Dancer from other species.

The Vivid Dancer is another western species found as far north as southern British Columbia and Alberta. It is common throughout Oregon with the exception of the extreme northwest corner of the state. It prefers springs and small spring-fed streams that are well-vegetated. It has been found at elevations up to 6,300 feet. Copulation takes place away from the water. The male remains attached to the female for several hours until tandem oviposition takes place (Corbet, 1999). Look for the Vivid Dancer perched on vegetation, rocks, logs, and the ground. It is one of the common damselflies in urban streams and drainage channels. We often see them perched on the sidewalks of residential neighborhoods in Eugene.

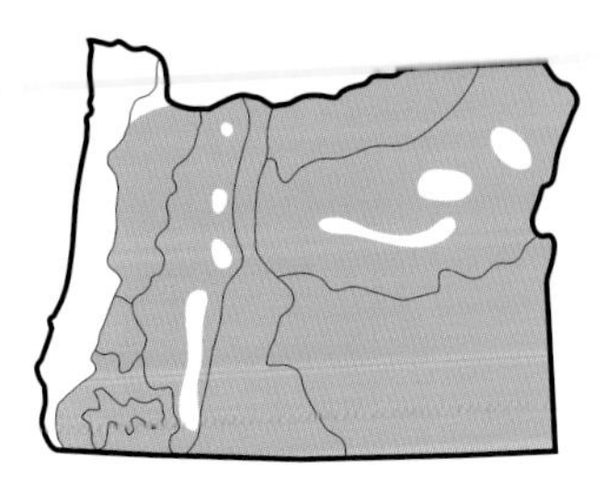

The adult flight season extends from early May to early October.

JAN	FEB	MAR	APR	MAY	JUN	JUL	AUG	SEP	OCT	NOV	DEC

Vivid Dancer male

Vivid Dancer female • photo by Jim Johnson • (inset: androchrome female)

Medium: 34–38 mm

Bluets, Genus *Coenagrion* and Genus *Enallagma*

The bluets are small sky blue and black damselflies that are abundant throughout Oregon in ponds, streams, wetlands, and alkali lakes. The males are mostly blue with black stripes on the top and shoulders of the thorax and black and blue on the abdomen (see Identification Chart Q). The top of the head has a blue occipital bar and spots. The abdominal patterns vary among species, sexes, and stages of development. Some species of bluets are not possible to identify without examining the male abdominal appendages in hand. The females can even be more difficult to identify in that females have more than one color form, usually tan or blue. It will be necessary to examine females in detail and compare morphological characters to reliably identify them.

Oregon has only a single species of bluets in the genus *Coenagrion*, the Taiga Bluet (see Figure 9 below). The members of this genus, of which there are only three species in North America, are typically known as the Eurasian bluets because most species are found in the Old World.

Six species of American bluets in the genus *Enallagma* are found in Oregon. Bluets can appear similar to some blue species of dancers (Genus *Argia*). The bluets perch in vegetation or floating moss rather than on rocks or the ground like the dancers. Bluets are generally smaller, and hold the wings lower over the abdomen than the dancers. In the hand, examine the spurs on the tibia of the leg (see Figure 7, page 245). Bluets have short spurs while dancers have spurs that are longer than the distance between the spurs. With experience, you will begin to identify these two genera in the field.

Since the bluets often perch low on emergent vegetation over the water, they can be overlooked. When they fly, they fly extremely close to the surface of the water or amongst emergent vegetation, making them difficult to net.

We have organized the bluets into three groups: 1) the unique Taiga Bluet, 2) the American bluets where the male's middle abdominal segments (3 through 5) are approximately half black and half blue on top (River, Tule, and Alkali), and 3) the American bluets where the male's middle abdominal segments are predominately blue on top (Familiar, Northern, and Boreal).

Figure 9 •
Taiga Bluet Abdominal Segment Mark
***Coenagrion resolutum* (top view)**

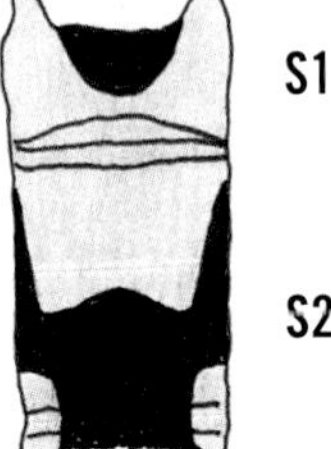

Distinctive black "U"-shaped mark on top of Segment 2

Identification Chart Q • Bluets, Genera: *Coenagrion* and *Enallagma*

Species (size)	Eye Spots (top view)	Abdomen (top view)	♂ Upper Appendages (side view)
Eurasian Bluets, Genus *Coenagrion* (Greenish undersides):			
Taiga *Coenagrion resolutum* Short 27–30 mm	 Mid-sized, "tear"-shaped	 S2 with "U"-shaped black mark (see Figure 9); S6 and S7 almost entirely black, diagnostic	 Long projectile with lobed tip
American Bluets, Genus *Enallagma*:			
Middle Abdominal Segments 3–5 half black, half blue			
River *Enallagma anna* Medium 30–36 mm	 Medium-sized	 Black areas pointed headward and increase in size from S2 to S7	 Long, extends beyond lower
Tule *E. carunculatum* Short 30–33 mm	 Large	Most segments half or more black	  Round, white tubercle
Alkali *E. clausum* Medium 33–35 mm	 Medium-sized	Black areas increase in size from S2 to S7	 Short, down-turned
Middle Abdominal Segments 3–5 mostly blue			
Familiar *E. civile* Long 35–39 mm	 Small, "tear"-shaped	 Black marks small, similar in size and shape	 Oval tubercle, black "shark tail"-shaped
Northern *E. annexum* Short 30–33 mm	 Large	S3–5 mostly blue	Hook on tip
Boreal *E. boreale* Short 33 mm	 Very large, almost touch eyes	  S3–5 mostly blue	 Rounded, blunt tip

Taiga Bluet

Coenagrion resolutum

The Taiga Bluet male has a blue upper thorax sometimes fading to green on the lower sides with the typical black stripe on the top and shoulders (see Identification Chart Q). The top of the blue second abdominal segment has a black "U"-shaped pattern (open end facing forward) (see Figure 9, page 258). This provides a good field identification mark. Abdominal segments 3 to 5 are about half blue and half black. Unlike the other bluets, segments 6 and 7 are black on top, but segments 8 and 9 are blue as in the other bluet species. These markings should allow for identification in the field with good binoculars. Taiga Bluet females have both green and blue color forms. Unlike other bluets, females of this species have no vulvar spine. The females and males are patterned similarly, but the females have more black on top of the abdomen, as do the other bluets. The male's upper appendages have a unique, long, downward angled projection with a lobed tip.

The Taiga Bluet is distributed across the northern United States and Canada, ranging to Alaska. The range of this bluet may extend further north than any other damselfly (Westfall and May, 2006). Thus, the name "Taiga" refers to the northern boreal forests. In Oregon, it is a mountain species of bogs and lake and pond margins above 3,000 feet in elevation.

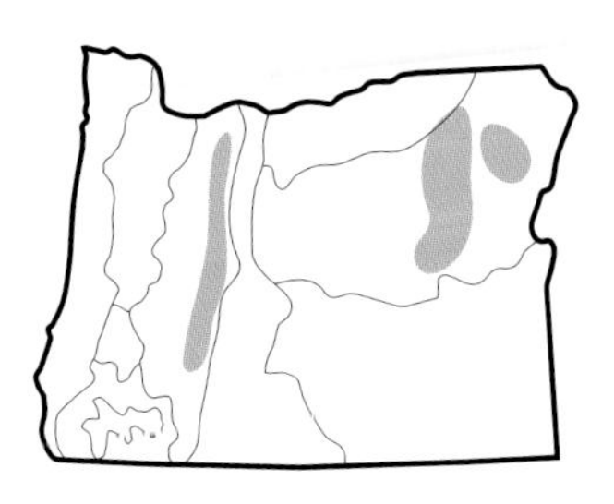

Perhaps due to its high-altitude habitat, the Taiga Bluet flies only from late May to late August in Oregon.

JAN	FEB	MAR	APR	MAY	JUN	JUL	AUG	SEP	OCT	NOV	DEC

Taiga Bluet male

Taiga Bluet female

Short: 27–33 mm

River Bluet

Enallagma anna

The River Bluet, along with the Tule and Alkali Bluets, make a group of three Oregon bluets whose middle abdominal segments are marked on top with about half black and half blue. Abdominal segments 2 through 7 are blue with black on top. The black spots on these segments gradually increase in size with each segment and have an arrow-like projection that all point forward toward the head. Abdominal segments 8 through 10 are blue with a dark area on top of segment 10. In side view, the male's upper appendages are unique with a rounded projection above, which extends far beyond the lower appendage. In fact, this feature can sometimes be seen without a hand lens. The River and Alkali Bluets are our largest bluet species.

The eminent Odonata researcher, E. B. Williamson, named this species for his future wife, Anna Tribolet. River Bluet females and males are patterned similarly, but the female's pale tan colors replace the blue found on the male. Abdominal segments 8 through 10 have a black stripe on top. There are also females with a pale blue coloration similar to the male's color.

As the name implies, the River Bluet is somewhat unusual for the bluets in that it inhabits flowing waters from about 3,300 to 5,800 feet in elevation. It can also be associated with the outflow from hot springs (Westfall and May, 2006). It is only known from 11 eastern Oregon counties, and we have not found it to be common. Outside Oregon, the River Bluet ranges from southern Alberta to New Mexico and east to Michigan.

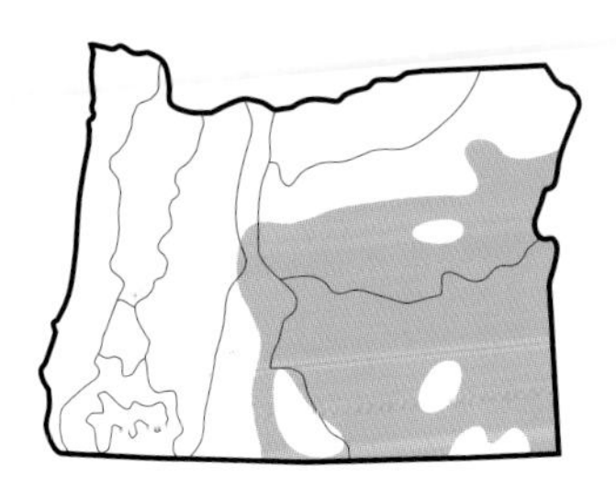

The adult flight season extends from early June to early October.

JAN	FEB	MAR	APR	MAY	JUN	JUL	AUG	SEP	OCT	NOV	DEC

River Bluet male

River Bluet androchrome female (inset: gynochrome female) • photos by Ray Bruun

Long: 30–36 mm

Northern Bluet

Enallagma annexum

The Northern Bluet is one of the three Oregon bluets (others are Familiar and Boreal) with the top of its abdomen more blue than black. The Northern and Boreal Bluets have almost identical field marks and can be told apart most easily by comparing their terminal appendages in hand (see Identification Chart R). The Northern Bluet is bright blue with middle abdominal segments 3 through 5 mostly blue above. The thoracic side stripe is undivided with a jag near the front. Segments 6 and 7 are mostly black, and segments 8 and 9 are blue. The eye spots on the back of the head are large and blue. In side view, the lower appendages are much longer than the upper appendages, and the upper appendage has a small upturned hook on the tip, a diagnostic feature. Northern Bluet females and males are patterned alike, but the female is tan with more black on the abdomen. The blue female form is patterned similarly, but the abdomen is mostly black with small amounts of blue. Females have been recorded crawling down plant stems and ovipositing up to a meter below the water surface (Westfall and May, 2006).

The Northern Bluet is found across North America and ranges from Alaska south into Mexico. It is common throughout Oregon up to 5,700 feet in elevation. It inhabits ponds, lakes, springs, and slower portions of streams and rivers. It prefers cooler waters.

The Northern Bluet's scientific name was changed to *Enallagma annexum* from *E. cyathigerum.*

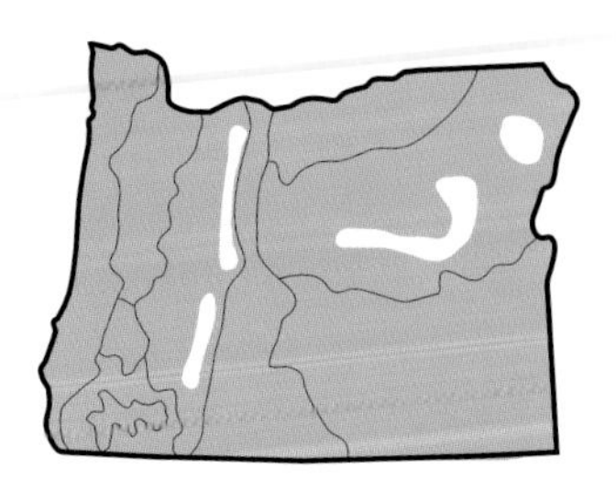

The adult flight season extends from mid-May to mid-October.

JAN	FEB	MAR	APR	MAY	JUN	JUL	AUG	SEP	OCT	NOV	DEC

Northern Bluet male

Northern Bluet female • photo by Ray Bruun

Medium: 30–33 mm

Boreal Bluet

Enallagma boreale

The Boreal Bluet is almost identical to the Northern Bluet, and differentiating them in the field is difficult. Along with the Northern and Familiar Bluets, the top of the Boreal Bluet's abdomen is mostly blue with small black markings. Its eye spots behind the head are very large and almost touch the eyes. However, the male's upper appendage is rounded and does not show the hook-like projection found in the Northern Bluet (see Identification Chart Q). One will have to examine specimens closely in hand to make a species determination. As with the Northern Bluet, this species has both tan and blue forms of the female with similar patterns. It is so common in its mountain habitats that you will see hundreds while examining bluets in hand—while looking for species other than Boreal Bluets.

The Boreal Bluet occurs as far north as the northern mountain slopes of the Yukon Territory near the Beaufort Sea (Westfall and May, 2006), indicating how resilient these insects are. As its name implies, this is a northern damselfly, which is very common in Oregon's mountainous ponds, lakes, and wetlands up to 7,500 feet in elevation. However, it also occurs on the coastal plain. This damselfly was found breeding in 2005 at Sandpiper Pond in Eugene, where it prefers emergent vegetation farthest from shore at the edge of the open water. This is one of the few known Willamette Valley lowland locations for this species.

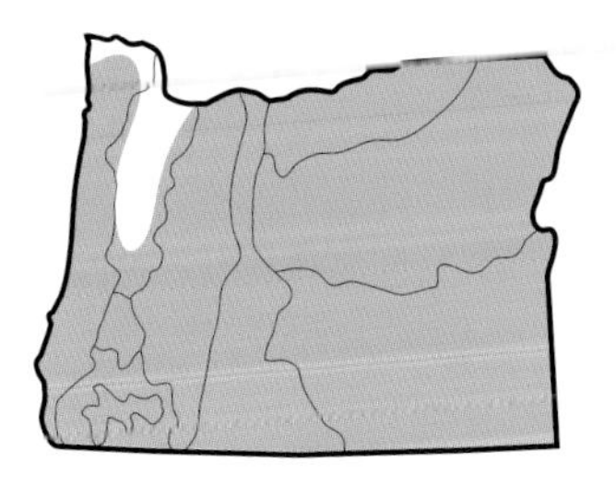

The adult flight season extends from early May to early November.

JAN FEB MAR APR MAY JUN JUL AUG SEP OCT NOV DEC

Boreal Bluet male

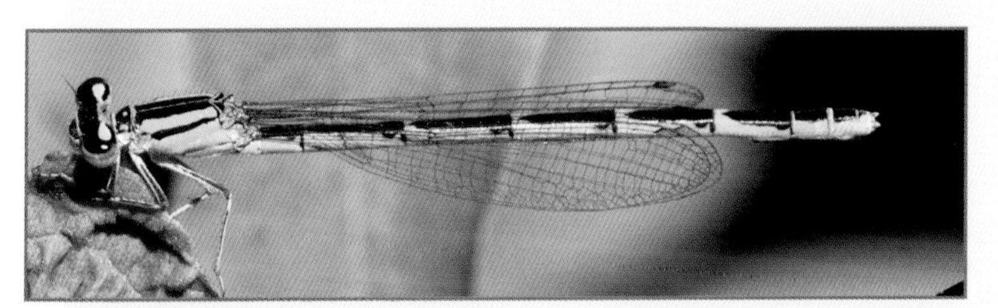

Boreal Bluet androchrome female

Medium: 31–35 mm

Boreal Bluet copulating pair with gynochrome female

Tule Bluet

Enallagma carunculatum

The Tule Bluet, along with the River and Alkali Bluets, make a group of three Oregon bluets whose middle abdominal segments are marked on top with about half black and half blue. Upon closer inspection you can see that the Tule Bluet has more black on the middle abdominal segments than does either the River or Alkali Bluets. The Tule Bluet has large blue eye spots on the back of the head. When viewed from the side, the male Tule Bluet's upper appendages are longer than the lower appendages. The tip of the upper appendage exhibits a round, whitish node, or caruncle, which is the source for its scientific name. Compare the shape of the upper appendage and caruncle with the Familiar Bluet (see Identification Chart Q). Both blue and tan form females of the Tule Bluet are known.

The Tule Bluet ranges from British Columbia south into Mexico and throughout the western and northeastern United States. The Tule Bluet is common throughout Oregon and can be found at a wide variety of ponds, lakes, and slow-moving streams with emergent vegetation up to about 2,500 feet in elevation. It is less common at higher elevations up to 7,500 feet. It is sometimes abundant, with hundreds along pond shorelines. In Eugene at Grimes Pond in summer, pairs in tandem can be seen swarming over floating algae and vegetation as females lay hundreds of eggs. In the low light of late afternoon, the sheer numbers of mating pairs multiplied by the magnitude of tiny eggs laid by each female help you realize why these insects survive through the ages; it is beautiful and marvelous to watch. Take time to observe the behaviors of this abundant species.

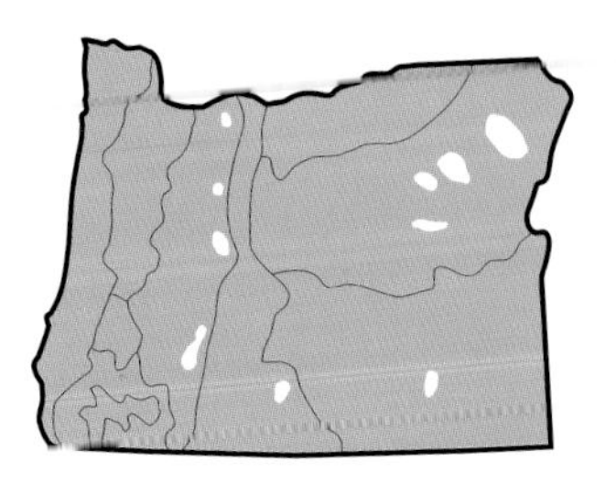

This Bluet has a very long flight season and can be found from early April to mid-November.

JAN FEB MAR APR MAY JUN JUL AUG SEP OCT NOV DEC

Tule Bluet male

Tule Bluet female

Short: 26–37 mm

Familiar Bluet

Enallagma civile

The large Familiar Bluet, along with the Northern and Boreal Bluets, is one of three Oregon bluets with the top of its abdominal segments predominately blue rather than black—but its black marks are smaller and more similar in shape than are those of the Northern or Boreal Bluets. It has very small, tear-shaped blue eye spots on the back of the head. The females have tan and blue forms which are patterned as other bluet females, similar to males but with more extensive black on the abdomen. Familiar Bluet females and juveniles can show a range in colors from pale blue, to rich blue, to lavender-gray (Manolis, 2003). In side view, the male's upper appendages are distinctive with a black fin-shaped form with a pale oval lobe on the trailing edge (see Identification Chart Q). The upper appendage extends further than the upturned hook on the lower appendage. Cases of hybridization between Familiar and Tule Bluets have been recorded.

The Familiar Bluet is found at lower elevations (up to 4,600 feet) in marshes and along creeks and rivers. It also colonizes temporary and human-created waters like irrigation ditches, stock ponds, reservoirs, and wastewater ponds. It forages over fields, meadows, and backyard gardens, and is known to steal prey from spider webs—and sometimes falling prey in the process (Manolis, 2003). The female oviposits in tandem with the male over water or in shoreline vegetation. In most cases, the female will become completely submerged during the oviposition sequence (Westfall and May, 2006).

The Familiar Bluet ranges from Manitoba, Canada south into South America, and it is widespread in the eastern and central United States. The Familiar Bluet is currently known from only five counties in southern and eastern Oregon. In California, it has extended its range northward (Manolis, 2003), and it may be expanding its range northward from Douglas, Josephine, and Jackson Counties in Oregon, as well . Since the Familiar Bluet is adept at exploiting human-created water features, it will likely continue expansion of its range in Oregon.

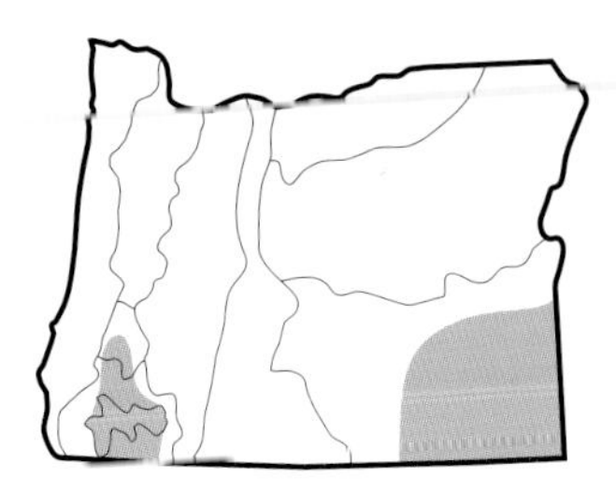

The adult flight season extends from late May to mid-October.

JAN	FEB	MAR	APR	MAY	JUN	JUL	AUG	SEP	OCT	NOV	DEC

Familiar Bluet male

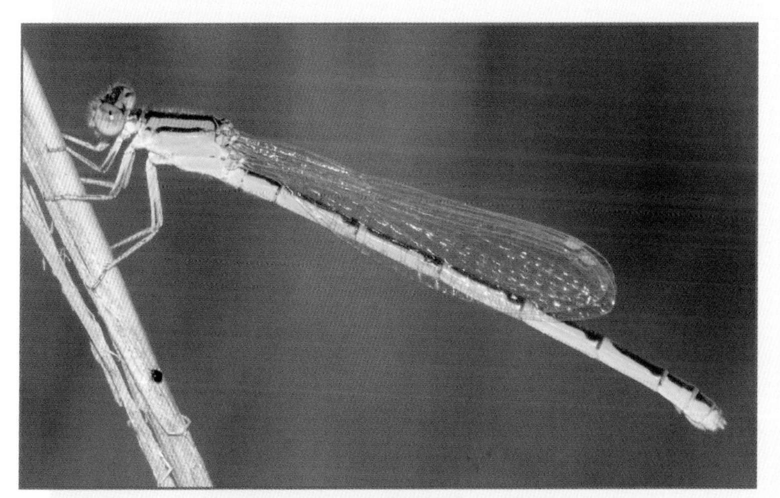

Familiar Bluet androchrome female • photo by Dennis Paulson

Long: 31–39 mm

Alkali Bluet

Enallagma clausum

The Alkali Bluet is one of three Oregon bluets with the top of its abdominal segments about half black and half blue—but it has a bit more blue on the middle segments than do the River or Tule Bluets. It is similar in size to the River Bluet, and these two species are larger than our other bluets. The abdominal segments are blue with areas of black on top. These dark areas are more prominent approaching the tip of the abdomen. The dorsal surface of S10 has a black stripe. In side view, the male's upper appendages are unique and are short, wedge-shaped, and pointed downward (see Identification Chart Q). Females can be either olive green or blue. The blue females and the male are patterned similarly, but the females have more black areas on top of the abdomen. The green female also has a similar pattern, but is a very pretty olive shade of green. We have also seen immature blue females that were a very pale blue, appearing almost gray.

This species occurs in alkaline lakes of eastern Oregon (thus the origin of its name), but it may also be found in fresh water. It is capable of surviving in salinity much greater than sea water (Westfall and May, 2006). In the eastern Oregon desert, it is often found perched up in the sagebrush higher than other bluets usually perch.

The Alkali Bluet's range extends from Saskatchewan south to southern New Mexico. In Oregon, it is most common in southeastern Oregon but has been found in Douglas and Deschutes Counties. It can be very common in the alkaline lakes that are its preferred habitat.

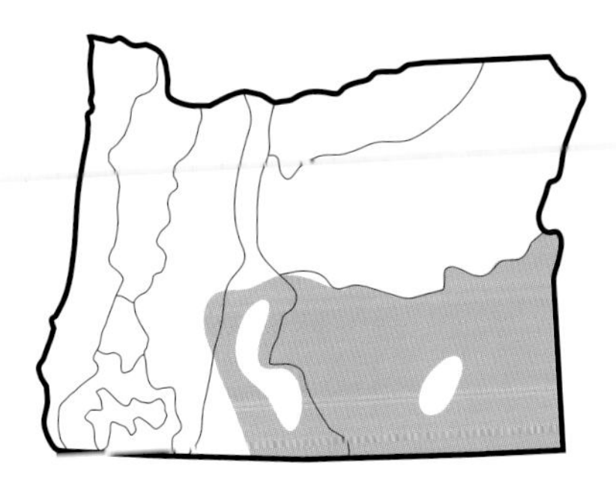

Its adult flight season extends from late May to late August.

JAN	FEB	MAR	APR	MAY	JUN	JUL	AUG	SEP	OCT	NOV	DEC
				▬	▬	▬	▬				

Alkali Bluet male

Alkali Bluet female (inset: androchrome female)

Long: 31–35 mm

Forktails, Genus *Ischnura*, and Sprites, Genus *Nehalennia*

We present these small damselflies in the same section for comparison purposes because they have similar field marks. These two genera have dark abdomens with blue tips, unlike the bluets and dancers.

Forktails, Genus *Ischnura*

Four of the fourteen North American species in this genus are found in Oregon. The forktails are small to medium-sized damselflies with a very slender abdomen, and the male's abdomen is completely black with a blue tip. Thus, although they are blue and black, they are easily separated from the bluets. The name "forktail" comes from the forked projection at the top of the end of the male's tenth abdominal segment (see Identification Chart R). The forktail females can be tan, orange, or blue. This group seems to have more female color forms than some other genera. The advantage of these female color forms has been postulated to be reduced interference from males during oviposition or mating for the androchrome (male color form), and reduced predation on the heterochrome (Westfall and May, 2006).

Females oviposit unattended by males with one exception in Oregon, the Black-fronted Forktail. Forktails are found in weeds along ponds and small streams. Although they seem to be weak fliers, they are certainly widely dispersed and common at many sites. The forktails feed both by gleaning and catching prey in flight (Corbet, 1999).

The forktails can be easily identified in the field with practice. The top of the thorax is a key feature to observe in the field. Identification Chart R illustrates the male forktail appendages in top and side views, as well as other key features.

Sprites, Genus *Nehalennia*

There are five species of sprites known from North America, but we have only a single species in Oregon. These are very small damselflies about an inch in length, usually having metallic green colors on the thorax (see Identification Chart R). They are usually found in dense vegetation in bogs and marshes. Sprites do a minimum of flying and are therefore easily overlooked in heavy vegetation. They oviposit in tandem.

Identification Chart R • Forktails, Genus *Ischnura*, and Sprite, Genus *Nehalennia*

Species (size)	Eye Spots (back of head)	Thorax (top view)	Abdomen (top view S8–10)	♂ Appendages (side and top views)
Forktails, Genus *Ischnura* (top S10 with prominent rearward extensions):			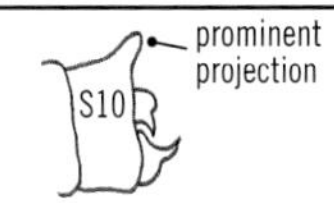	
Swift ***Ischnura erratica*** Long 31–34 mm	 "Tear"-shaped, blue	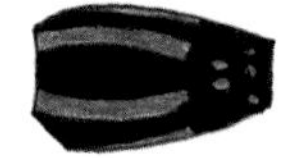 Black with broad, blue stripes	 S8 and S9: blue, "urn"-shaped marks; sometimes S7 "urn" mark, diagnostic	 upper: Long, hooked downward; lower: Long, thin, hooked upward
Western ***I. perparva*** Short 25–27 mm	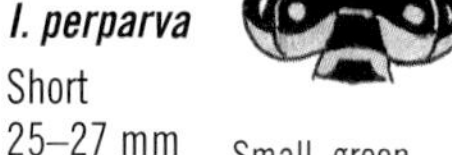  Small, green	 Black with green stripes; greenish sides	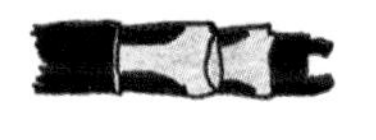 S8 and S9 black marks on basal edges	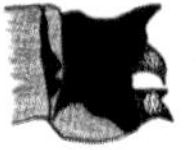 lower: Forked
Pacific ***I. cervula*** Medium 25–30 mm	 Tiny, blue	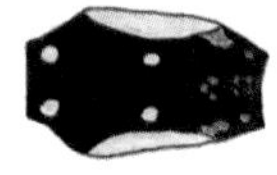 Black with 4 pale blue spots, diagnostic	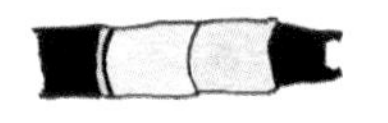 S8 and S9 blue, no black marks	 upper: Short, hooked downward; lower: Long, hooked upward
Black-fronted ***I. denticollis*** Short 20–26 mm	 Tiny, blue or green	Solid iridescent black	S8 "pawn"-shaped blue mark; S9 "urn"-shaped blue mark	 upper: Hooked downward; lower: Thin, short
Sprite, Genus *Nehalennia* (S10 with serrated rear edge): 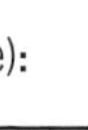				
Sedge ***Nehalennia irene*** Short 24–29 mm 	None, diagnostic	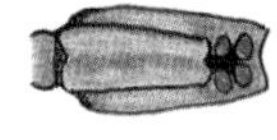 Metallic green	 S9 and S10 black marks at base; S8 variable amount of blue	 Short, stubby

Pacific Forktail

Ischnura cervula

The thorax is blue on the sides and black on the top, with a distinctive blue dot at each of the four corners. This feature can be seen with the unaided eye or with binoculars and makes this species very easy to identify in the field. The top of the abdomen is a typical forktail black, while segments 8 and 9 are extensively blue. The eye spots at the back of the head are tiny and blue. We have observed three color forms of female Pacific Forktails. The gynochrome is a tan color with a blue abdominal tip. One blue form of the female has elongated marks where the male's front spots are located, thus giving it an exclamation mark pattern of blue on the black thoracic top. Finally, there is a blue form that has a solid blue stripe indented in the center in place of the dots, giving it a thoracic pattern similar to the bluets.

The Pacific Forktail is found in western North America from British Columbia into Mexico and is very common throughout Oregon up to 7,300 feet in elevation. It is also one of our earliest damselflies to appear in the spring. We have observed them locally as early as late January. At Delta Ponds in Eugene during late spring and summer, when the Pacific Forktails are in full breeding frenzy, they will land on your cap, arms, hands, and net. You may also see all of the female color forms during this time. In those circumstances, you won't need the lens or binoculars to study them first hand.

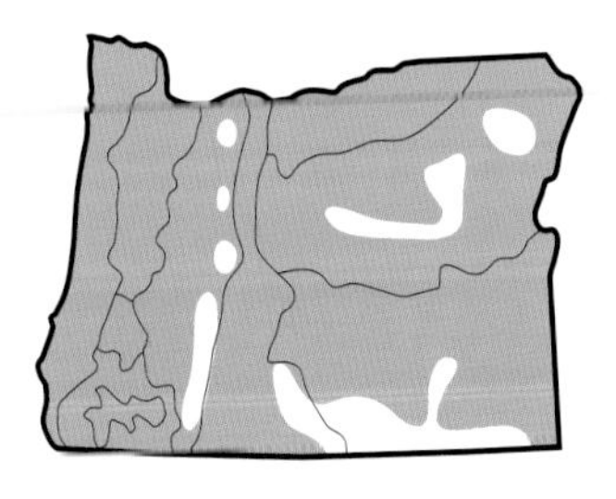

The adult flight season extends from late January to the end of October.

JAN FEB MAR APR MAY JUN JUL AUG SEP OCT NOV DEC

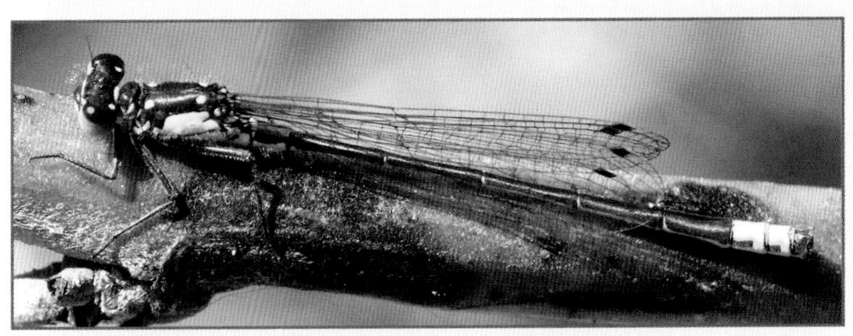

Pacific Forktail male

Pacific Forktail gynochrome female

Pacific Forktail androchrome female

Pacific Forktail copulating pair with androchrome female

Black-fronted Forktail

Ischnura denticollis

The very small Black-fronted Forktail is appropriately named, having a completely black top on the thorax. The sides of the thorax are blue, with blue on the bottom of the black abdominal segments. The top of abdominal segments 8 and 9 have blue wedge-shaped patches. The blue or green spots behind the eye are very small. The female is patterned similarly to the male, but a tan or green color replaces the blue. The black on the top of the female thorax is reduced, with a pale stripe on each side separated by a narrow black stripe from the tan on the sides of the thorax. The tip of the abdomen retains the typical forktail coloration with light blue patterns on segments 8 and 9. In addition, there is a female color form that resembles the male and may have blue shoulder stripes. In side view, the male's upper appendages are hooked downward and the lower appendages are thin, short, and curved upward.

Oregon is the northernmost distribution of the Black-fronted Forktail, which has a southwestern range extending south into Mexico and east into northeastern Kansas. This forktail is mainly found in the vegetated margins of hot springs in southeastern Oregon up to 4,500 feet in elevation. However, Jim Johnson found it at a pond in White City, Jackson County.

This is one of only two forktails in North America that oviposit in tandem. The male of the Black-fronted Forktail has a scoop on the penis which may be used to remove sperm from prior matings. Thus, guarding of the female may have evolved to prevent this from being accomplished by competing males (Corbet, 1999).

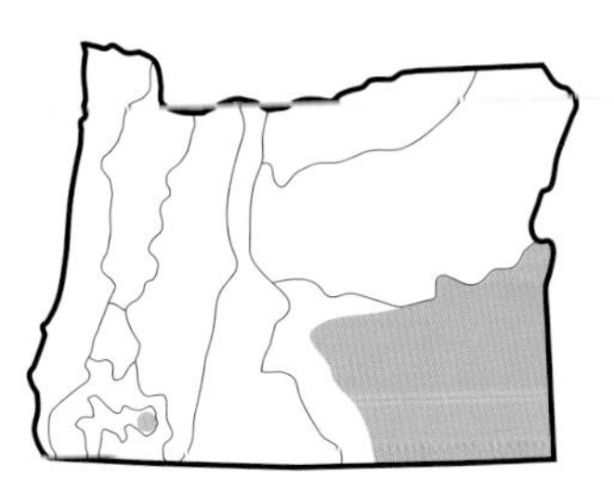

The Black-fronted Forktail's adult flight season extends from late April into mid-October.

JAN	FEB	MAR	APR	MAY	JUN	JUL	AUG	SEP	OCT	NOV	DEC

Black-fronted Forktail male

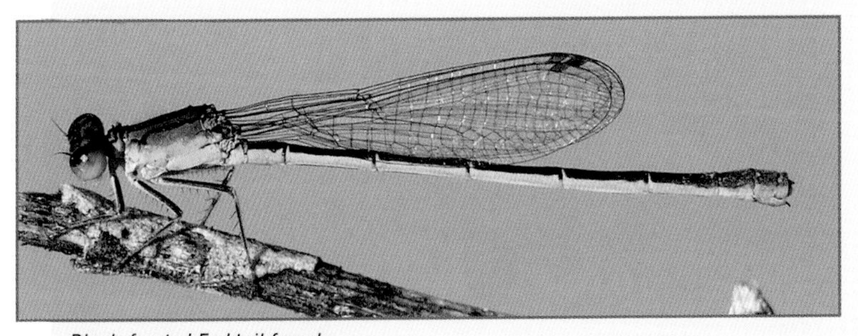

Black-fronted Forktail female

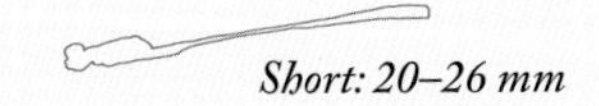

Short: 20–26 mm

Black-fronted Forktail copulating pair

Swift Forktail

Ischnura erratica

The Swift Forktail is our largest forktail. The thorax has blue sides and is black on top with two broad, lateral, deep blue stripes. The tip of the abdomen has more extensive blue than do the Pacific or Western Forktails, with the blue on segments 8 and 9 extending to segment 10 and most often extending forward onto segment 7. Segments 8 and 9 have "urn"-shaped blue patterns on top. Segment 10 has an elongated tip that forks upward and the lower appendage is long with an upward hook, giving the Swift Forktail an accentuated "forked" tail appearance (see Identification Chart R). The long lower appendage of the male can be seen with binoculars allowing for field identification. The female may be colored like the male or have a green thorax, and teneral young females may have an orange thorax.

The Swift Forktail's range is restricted to the west coast from southern British Columbia to northern California. It is found in Oregon from the coast to just east of the Cascade crest at about 4,800 feet in elevation. It has been found at two locations in Oregon's Blue Mountains and should be looked for at suitable habitat. The Swift Forktail is a robust flier that frequents ponds, especially beaver ponds, lakes, and river floodplains with clear, shallow water and emergent vegetation. We have also found it at clear streams such as Salt Creek at Gold Lake in Lane County and Crescent Creek in Klamath County in backwater habitats. Salt Creek is a good spot to see this species in July. Later in the season, it can be difficult to find. The female lays her eggs unattended by the male. This damselfly often perches in open areas and escapes up into a tree or bush, so it likes water with some woody fringe.

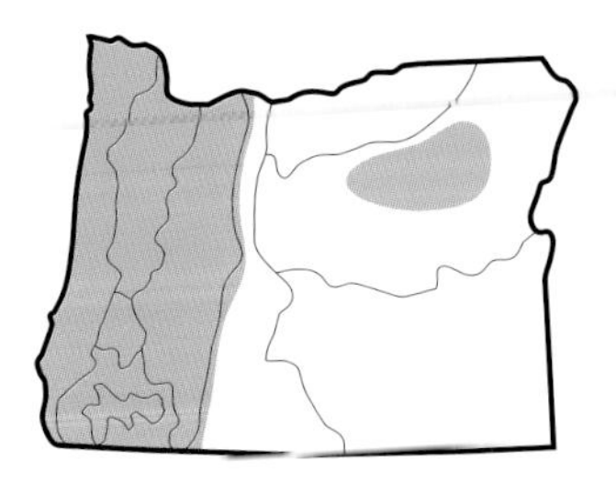

The Swift Forktail is an early-season species with an adult flight season extending from mid-March to late August.

JAN	FEB	MAR	APR	MAY	JUN	JUL	AUG	SEP	OCT	NOV	DEC

Swift Forktail male

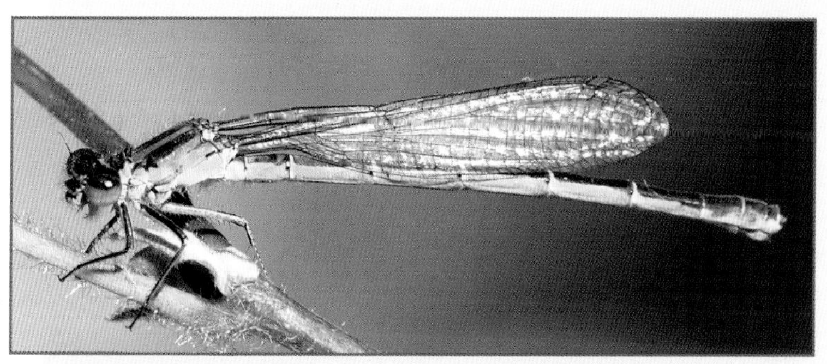

Swift Forktail gynochrome female

Swift Forktail androchrome female

Long: 31–34 mm

Western Forktail

Ischnura perparva

This is one of our smallest Odonates. The thorax has pale green sides, and the top of the thorax is striped with black and blue or bluish-green. The underside is pale green. It is our only forktail with a forked lower appendage (see Identification Chart R). There is extensive blue on the back of the head with two, small, greenish eye spots. The blue segments 8 and 9 are edged with black marks, which extend rearward from the base of each of those segments. The wings on the Western Forktail appear to be very short. The mature female has a dark thorax and a blue abdomen, both covered with a bluish pruinosity. This unique coloration makes field identification easy. The eyes are dark above and green below. With experience, both males and females may be easily identified in the field.

This species is found in western North America from British Columbia into Mexico and east into the Dakotas and Nebraska. It is fairly common in Oregon up to 7,800 feet in elevation at marshy ponds, lakes, and slow streams. In Lane County, it has been found frequently at Grimes and Sandpiper Ponds in the west Eugene wetlands.

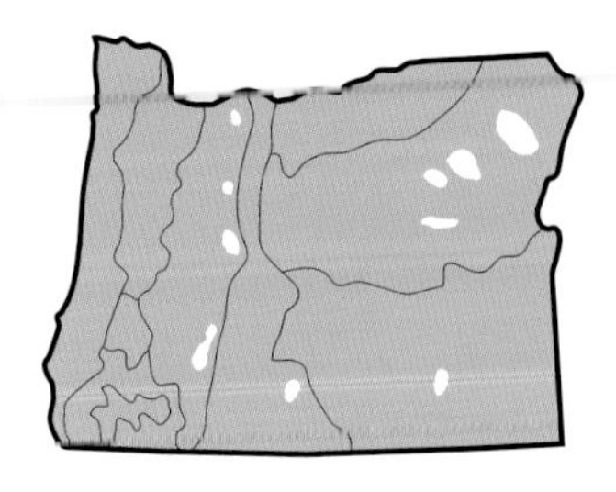

The adult flight season extends from early April into late October.

JAN	FEB	MAR	APR	MAY	JUN	JUL	AUG	SEP	OCT	NOV	DEC

Western Forktail male

Western Forktail female (inset: side view)

Medium: 25–30 mm

Sedge Sprite

Nehalennia irene

The small male Sedge Sprite is metallic green on the top of the head and top and upper sides of the thorax. The eyes are blue and the lower side of the thorax is light blue. The top of the abdomen is metallic green but can appear dark depending upon lighting. Abdominal segments 8 and 9 are powder blue on the posterior with basal black marks that form a blue "W" shape on each of those segments. Unlike the small forktails, the sprite's segment 10 is blue rather than black, it lacks eye spots on the back of its head, and it lacks the projection at the top end of abdominal segment ten; instead it has a serrated edge. The bottom of the abdomen is white to cream color. Sedge Sprite males and females are similar in color and pattern. It is easily identified in the field and should not be confused with any other Oregon species. The male's upper and lower appendages are short and stubby.

This small damselfly is easily overlooked, but it is well worth the effort to look for it. Its small size masks its truly spectacular colors but close-focusing binoculars and its approachability can provide you with a great view. If this little damselfly were larger, it would draw a lot of attention! As its name implies, it is often associated with sedge habitats in bogs and marshy areas.

The Sedge Sprite is found across the United States and Canada as far north as Alaska, but is currently known in Oregon only from Cascade crest sites in Lane, Jackson, Douglas, and Klamath counties at elevations above 4,500 feet.

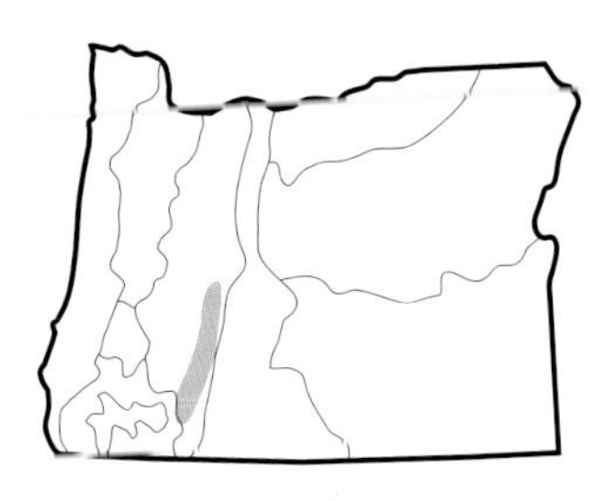

It has a short flight season in Oregon, and adults have only been found from late June to late August.

JAN	FEB	MAR	APR	MAY	JUN	JUL	AUG	SEP	OCT	NOV	DEC

Sedge Sprite male

Sedge Sprite copulating pair

Medium: 24–29 mm

Red Damsels, Genus *Amphiagrion*

The two North American species in this genus are named for their red coloration. The lone Oregon representative, the Western Red Damsel, *Amphiagrion abbreviatum*, is named for its relatively short abdomen. Besides its red color and short abdomen, this species displays two other interesting characteristics: a) a large tubercle on the underside of the thorax at the rear, and b) a pterostigma that is longer on the leading edge than its back edge, giving it a rhomboid appearance. The female has a strong vulvar spine.

Western Red Damsel

Amphiagrion abbreviatum

The male has a dark head and a black upper thorax, which are unusually hairy. The stocky abdomen is bright red with small black spots at the rear on the sides of abdominal segments which become progressively larger on segments 6 through 10. The legs are light colored with a black stripe on the femora. The female's body color is a light orange color with a reddish blush on the top of the abdomen. The small size and unique color of the Western Red Damsel make it easy to identify in the field. At rest, the abdomen barely extends beyond the folded wing tips, thus the scientific name, *abbreviatum*, for the stubby abdomen. In the hand, note the conspicuous bump on the underside of the thorax. It is difficult to appreciate the wonderful colors of this damsel due to its very small size and its tendency to stay among emergent vegetation.

This is a western species found from southern British Columbia and Alberta to southern California and New Mexico. The Western Red Damsel favors still waters in ponds, hot springs, ditches, slow moving creeks, and adjacent meadows, particularly where sedges are present . This small damsel's rapid flight close to the water and in vegetation makes it easy to overlook. It occurs at a wide variety of elevations and is found throughout Oregon from the Willamette Valley and up to 7,500 feet in elevation in the high Cascades. It is especially common in Cascade Mountain bogs and wet meadows. This species occurs throughout the Willamette Valley and is locally common. Mating pairs oviposit in floating vegetation.

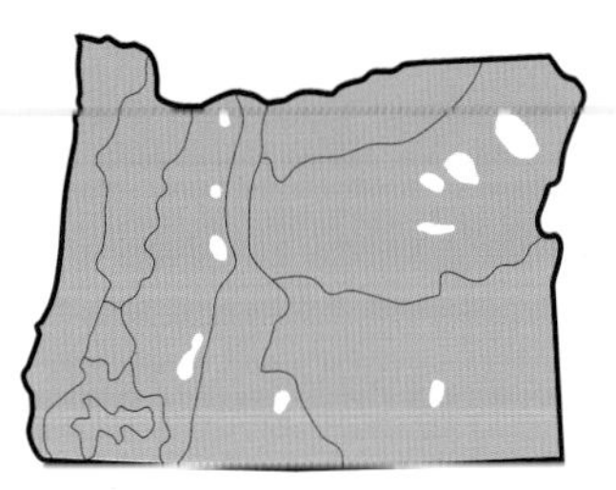

This damselfly tends to be more common early in the flight season, which extends from late April to late August.

JAN	FEB	MAR	APR	MAY	JUN	JUL	AUG	SEP	OCT	NOV	DEC

Western Red Damsel male

Western Red Damsel female

Short: 25–28 mm

Common Green Darner ovipositing

The Wonder of Dragonflies

Appendix A.

Oregon Dragonfly and Damselfly Checklist (91 species total)

Dragonflies (Anisoptera) (63 species)

Petaltails, Petaluridae

Black Petaltail, *Tanypteryx hageni*

Darners, Aeshnidae

Canada Darner, *Aeshna canadensis*
Lance-tipped Darner, *Aeshna constricta*
Variable Darner, *Aeshna interrupta*
Sedge Darner, *Aeshna juncea*
Paddle-tailed Darner, *Aeshna palmata*
Zigzag Darner, *Aeshna sitchensis*
Subarctic Darner, *Aeshna subarctica*
Black-tipped Darner, *Aeshna tuberculifera*
Shadow Darner, *Aeshna umbrosa*
Walker's Darner, *Aeshna walkeri*
Common Green Darner, *Anax junius*
California Darner, *Rhionaeschna californica*
Blue-eyed Darner, *Rhionaeschna multicolor*

Clubtails, Gomphidae

White-belted Ringtail, *Erpetogomphus compositus*
Pacific Clubtail, *Gomphus kurilis*
Columbia Clubtail, *Gomphus lynnae*
Grappletail, *Octogomphus specularis*
Bison Snaketail, *Ophiogomphus bison*
Great Basin Snaketail, *Ophiogomphus morrisoni*
Sinuous Snaketail, *Ophiogomphus occidentis*
Pale Snaketail, *Ophiogomphus severus*
Olive Clubtail, *Stylurus olivaceus*

Spiketails, Cordulegastridae

Pacific Spiketail, *Cordulegaster dorsalis*

Cruisers, Macromiidae

Western River Cruiser, *Macromia magnifica*

Emeralds, Corduliidae

American Emerald, *Cordulia shurtleffii*
Beaverpond Baskettail, *Epitheca canis*
Spiny Baskettail, *Epitheca spinigera*

Ringed Emerald, *Somatochlora albicincta*
Ocellated Emerald, *Somatochlora minor*
Mountain Emerald, *Somatochlora semicircularis*
Brush-tipped Emerald, *Somatochlora walshii*

Skimmers, Libellulidae
Western Pondhawk, *Erythemis collocata*
Chalk-fronted Corporal, *Ladona julia*
Crimson-ringed Whiteface, *Leucorrhinia glacialis*
Hudsonian Whiteface, *Leucorrhinia hudsonica*
Dot-tailed Whiteface, *Leucorrhinia intacta*
Belted Whiteface, *Leucorrhinia proxima*
Comanche Skimmer, *Libellula comanche*
Bleached Skimmer, *Libellula composita*
Eight-spotted Skimmer, *Libellula forensis*
Widow Skimmer, *Libellula luctuosa*
Hoary Skimmer, *Libellula nodisticta*
Twelve-spotted Skimmer, *Libellula pulchella*
Four-spotted Skimmer, *Libellula quadrimaculata*
Flame Skimmer, *Libellula saturata*
Blue Dasher, *Pachydiplax longipennis*
Red Rock Skimmer, *Paltothemis lineatipes*
Wandering Glider, *Pantala flavescens*
Spot-winged Glider, *Pantala hymenaea*
Common Whitetail, *Plathemis lydia*
Desert Whitetail, *Plathemis subornata*
Variegated Meadowhawk, *Sympetrum corruptum*
Saffron-winged Meadowhawk, *Sympetrum costiferum*
Black Meadowhawk, *Sympetrum danae*
Cardinal Meadowhawk, *Sympetrum illotum*
Cherry-faced Meadowhawk, *Sympetrum internum*
Red-veined Meadowhawk, *Sympetrum madidum*
White-faced Meadowhawk, *Sympetrum obtrusum*
Striped Meadowhawk, *Sympetrum pallipes*
Band-winged Meadowhawk, *Sympetrum semicinctum*
Autumn Meadowhawk, *Sympetrum vicinum*
Black Saddlebags, *Tramea lacerata*

Damselflies (Zygoptera) (28 species)

Broad-winged, Calopterygidae

River Jewelwing, *Calopteryx aequabilis*
American Rubyspot, *Hetaerina americana*

Spreadwings, Lestidae

California Spreadwing, *Archilestes californica*
Spotted Spreadwing, *Lestes congener*
Northern Spreadwing, *Lestes disjunctus*
Emerald Spreadwing, *Lestes dryas*
Sweetflag Spreadwing, *Lestes forcipatus*
Black Spreadwing, *Lestes stultus*
Lyre-tipped Spreadwing, *Lestes unguiculatus*

Pond Damsels, Coenagrionidae

Western Red Damsel, *Amphiagrion abbreviatum*
California Dancer, *Argia agrioides*
Paiute Dancer, *Argia alberta*
Emma's Dancer, *Argia emma*
Sooty Dancer, *Argia lugens*
Aztec Dancer, *Argia nahuana*
Vivid Dancer, *Argia vivida*
Taiga Bluet, *Coenagrion resolutum*
River Bluet, *Enallagma anna*
Northern Bluet, *Enallagma annexum*
Boreal Bluet, *Enallagma boreale*
Tule Bluet, *Enallagma carunculatum*
Familiar Bluet, *Enallagma civile*
Alkali Bluet, *Enallagma clausum*
Pacific Forktail, *Ischnura cervula*
Black-fronted Forktail, *Ischnura denticollis*
Swift Forktail, *Ischnura erratica*
Western Forktail, *Ischnura perparva*
Sedge Sprite, *Nehalennia irene*

Appendix B. Adult Flight Season Chart

Dragonflies

Petaltails, Petaluridae

Black Petaltail, *Tanypteryx hageni*

Darners, Aeshnidae

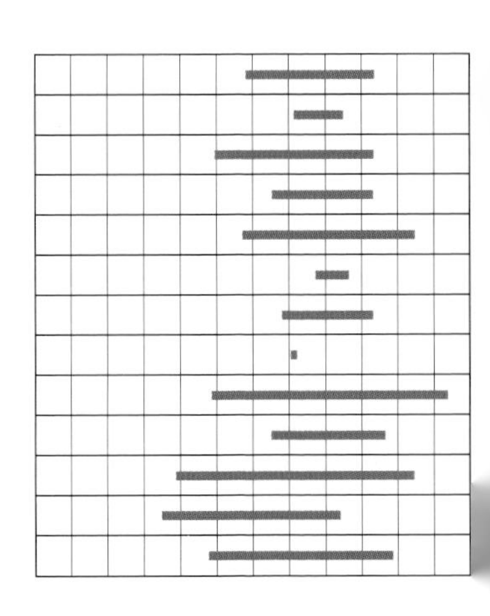

Canada Darner, *Aeshna canadensis*
Lance-tipped Darner, *Aeshna constricta*
Variable Darner, *Aeshna interrupta*
Sedge Darner, *Aeshna juncea*
Paddle-tailed Darner, *Aeshna palmata*
Zigzag Darner, *Aeshna sitchensis*
Subarctic Darner, *Aeshna subarctica*
Black-tipped Darner, *Aeshna tuberculifera*
Shadow Darner, *Aeshna umbrosa*
Walker's Darner, *Aeshna walkeri*
Common Green Darner, *Anax junius*
California Darner, *Rhionaeschna californica*
Blue-eyed Darner, *Rhionaeschna multicolor*

Clubtails, Gomphidae

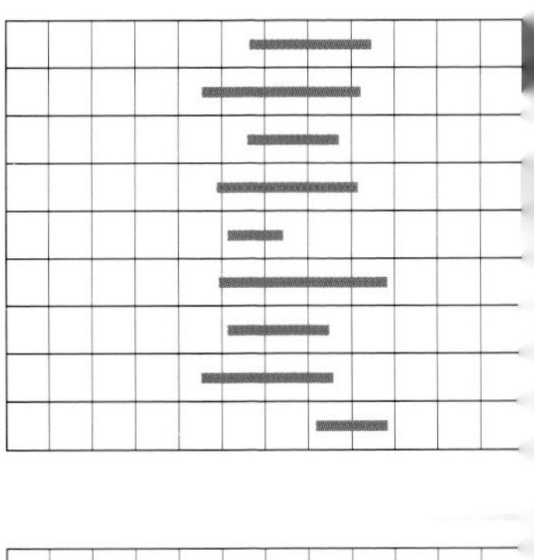

White-belted Ringtail, *Erpetogomphus compositus*
Pacific Clubtail, *Gomphus kurilis*
Columbia Clubtail, *Gomphus lynnae*
Grappletail, *Octogomphus specularis*
Bison Snaketail, *Ophiogomphus bison*
Great Basin Snaketail, *Ophiogomphus morrisoni*
Sinuous Snaketail, *Ophiogomphus occidentis*
Pale Snaketail, *Ophiogomphus severus*
Olive Clubtail, *Stylurus olivaceus*

Spiketails, Cordulegastridae

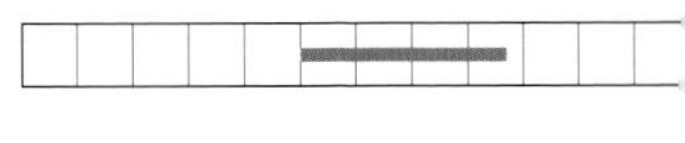

Pacific Spiketail, *Cordulegaster dorsalis*

Cruisers, Macromiidae

Western River Cruiser, *Macromia magnifica*

DRAGONFLIES (continued)

JAN FEB MAR APR MAY JUN JUL AUG SEP OCT NOV DEC

Emeralds, Corduliidae

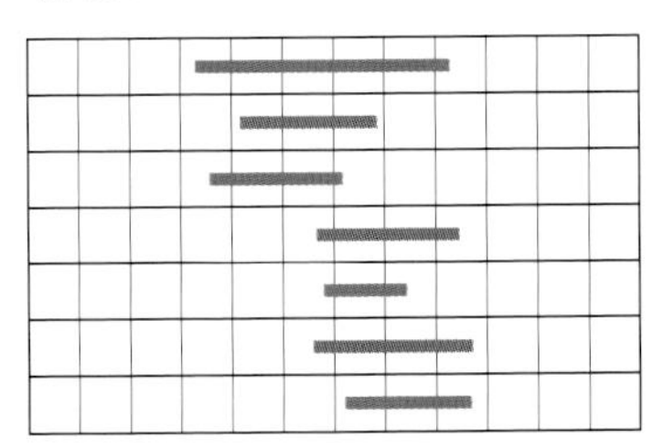

American Emerald, *Cordulia shurtleffii*
Beaverpond Baskettail, *Epitheca canis*
Spiny Baskettail, *Epitheca spinigera*
Ringed Emerald, *Somatochlora albicincta*
Ocellated Emerald, *Somatochlora minor*
Mountain Emerald, *Somatochlora semicircularis*
Brush-tipped Emerald, *Somatochlora walshii*

Skimmers, Libellulidae

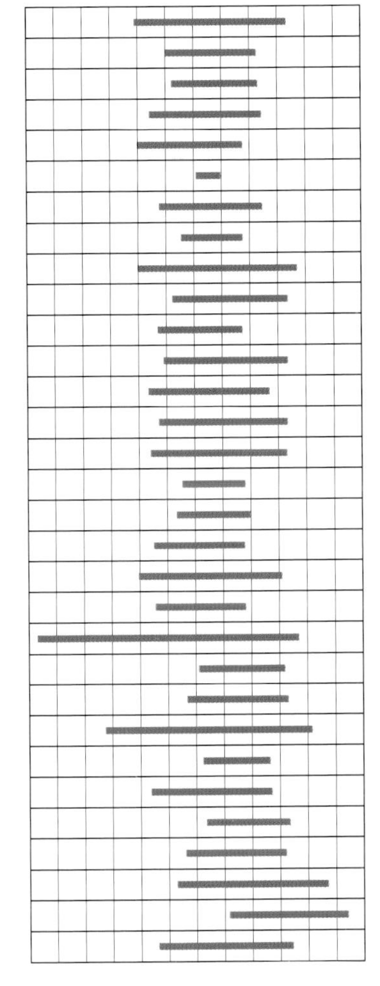

Western Pondhawk, *Erythemis collocata*
Chalk-fronted Corporal, *Ladona julia*
Crimson-ringed Whiteface, *Leucorrhinia glacialis*
Hudsonian Whiteface, *Leucorrhinia hudsonica*
Dot-tailed Whiteface, *Leucorrhinia intacta*
Belted Whiteface, *Leucorrhinia proxima*
Comanche Skimmer, *Libellula comanche*
Bleached Skimmer, *Libellula composita*
Eight-spotted Skimmer, *Libellula forensis*
Widow Skimmer, *Libellula luctuosa*
Hoary Skimmer, *Libellula nodisticta*
Twelve-spotted Skimmer, *Libellula pulchella*
Four-spotted Skimmer, *Libellula quadrimaculata*
Flame Skimmer, *Libellula saturata*
Blue Dasher, *Pachydiplax longipennis*
Red Rock Skimmer, *Paltothemis lineatipes*
Wandering Glider, *Pantala flavescens*
Spot-winged Glider, *Pantala hymenaea*
Common Whitetail, *Plathemis lydia*
Desert Whitetail, *Plathemis subornata*
Variegated Meadowhawk, *Sympetrum corruptum*
Saffron-winged Meadowhawk, *Sympetrum costiferum*
Black Meadowhawk, *Sympetrum danae*
Cardinal Meadowhawk, *Sympetrum illotum*
Cherry-faced Meadowhawk, *Sympetrum internum*
Red-veined Meadowhawk, *Sympetrum madidum*
White-faced Meadowhawk, *Sympetrum obtrusum*
Band-winged Meadowhawk, *Sympetrum semicinctum*
Striped Meadowhawk, *Sympetrum pallipes*
Autumn Meadowhawk, *Sympetrum vicinum*
Black Saddlebags, *Tramea lacerata*

Damselflies

Broad-winged Damsels, Calopterygidae

River Jewelwing, *Calopteryx aequabilis*
American Rubyspot, *Hetaerina americana*

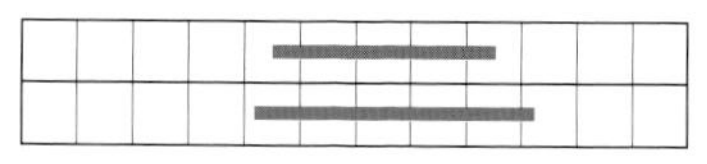

Spreadwings, Lestidae

California Spreadwing, *Archilestes californica*
Spotted Spreadwing, *Lestes congener*
Northern Spreadwing, *Lestes disjunctus*
Emerald Spreadwing, *Lestes dryas*
Sweetflag Spreadwing, *Lestes forcipatus*
Black Spreadwing, *Lestes stultus*
Lyre-tipped Spreadwing, *Lestes unguiculatus*

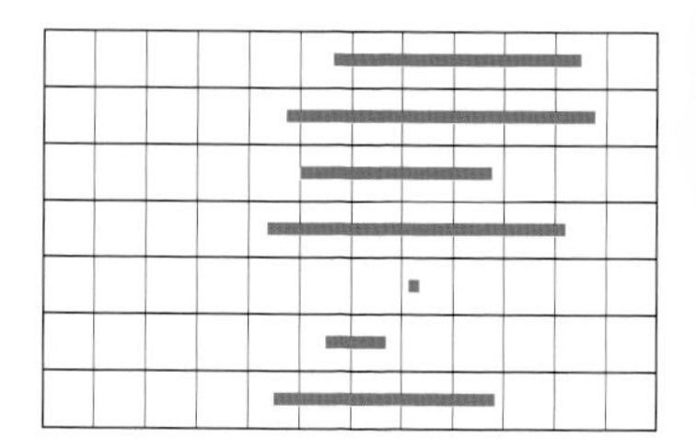

Pond Damsels, Coenagrionidae

Western Red Damsel, *Amphiagrion abbreviatum*
California Dancer, *Argia agrioides*
Paiute Dancer, *Argia alberta*
Emma's Dancer, *Argia emma*
Sooty Dancer, *Argia lugens*
Aztec Dancer, *Argia nahuana*
Vivid Dancer, *Argia vivida*
Taiga Bluet, *Coenagrion resolutum*
River Bluet, *Enallagma anna*
Northern Bluet, *Enallagma annexum*
Boreal Bluet, *Enallagma boreale*
Tule Bluet, *Enallagma carunculatum*
Familiar Bluet, *Enallagma civile*
Alkali Bluet, *Enallagma clausum*
Pacific Forktail, *Ischnura cervula*
Black-fronted Forktail, *Ischnura denticollis*
Swift Forktail, *Ischnura erratica*
Western Forktail, *Ischnura perparva*
Sedge Sprite, *Nehalennia irene*

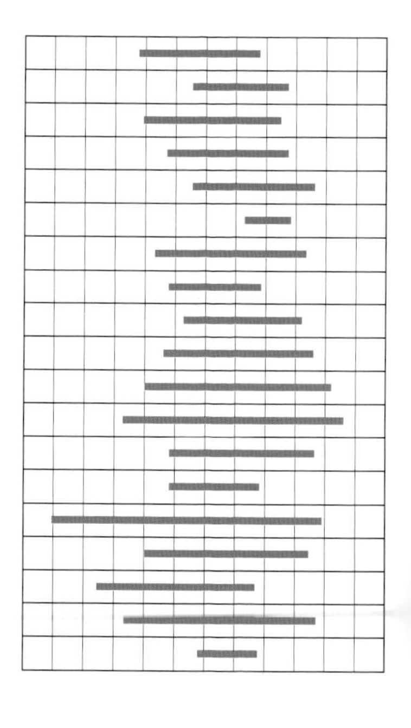

Appendix C.

The following table and graph depict the number of species of Odonata recorded flying in Oregon by month. The table and graph illustrate that few species have been recorded during the December to February winter months. They also show the July and August summer peak when eighty-six (95 %) of Oregon's ninety-one species have been recorded.

Figure 10 •
Oregon Odonate Abundance by Month
(Number of Adult Species)

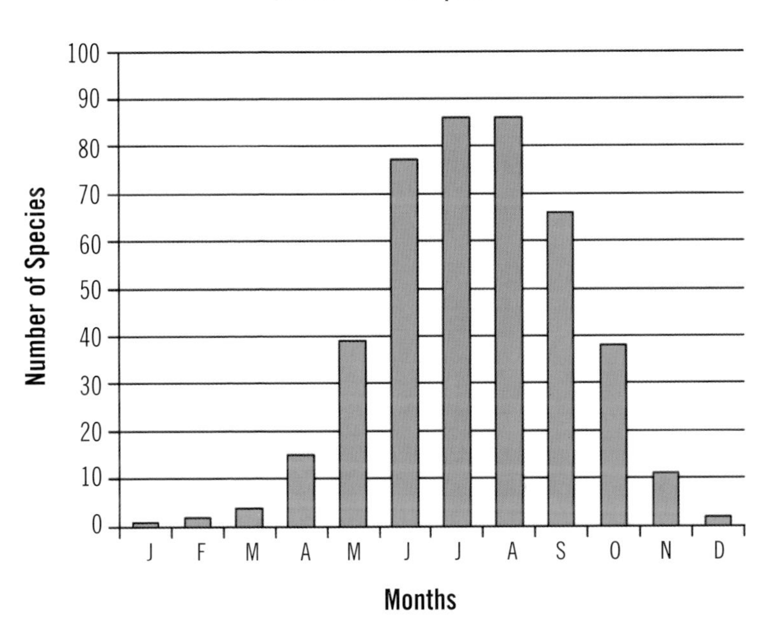

	JAN	FEB	MAR	APR	MAY	JUN	JUL	AUG	SEP	OCT	NOV	DEC
Number of Species	1	2	4	15	39	77	86	86	66	38	11	2

D. Name Changes

The Checklist Committee (formerly the Names Committee) of the Dragonfly Society of the Americas developed the list of English names for our Odonates in 1996, thus assisting many amateurs who wish to enjoy identifying these insects using English names. Both scientific and common names have changed occasionally with advances in our understanding of taxonomy. As scientific knowledge advances, future classifications and names will also change. We have tried to use the most current names (Paulson and Dunkle, 2009). This list will help you compare names used in this book with other modern field guides and references.

California Darner, *Rhionaeschna californica*........................... was *Aeshna californica*.

Blue-eyed Darner, *Rhionaeschna multicolor*............................ was *Aeshna multicolor*.

Beaverpond Baskettail, *Epitheca canis* .... was *Tetragoneuria canis*.

Spiny Baskettail, *Epitheca spinigera*.............................. was *Tetragoneuria spinigera*.

Chalk-fronted Corporal, *Ladona julia* .... was *Libellula julia*.

Belted Whiteface, *Leucorrhinia proxima* .... was Red-waisted Whiteface.

Common Whitetail, *Plathemis lydia* ....was *Libellula lydia*.

Desert Whitetail, *Plathemis subornata*.................................was *Libellula subornata*.

Band-winged Meadowhawk, *Sympetrum semicinctum*...
....was Western Meadowhawk *S. occidentale*.
(It was lumped with the Band-winged Meadowhawk, *S. semicinctum*)

Autumn Meadowhawk, *Sympetrum vicinum*.......was Yellow-legged Meadowhawk.

Northern Spreadwing, *Lestes disjunctus*..........................was Common Spreadwing.

Northern Bluet, *Enallagma annexum*...........................was *Enallagma cyathigerum*.

References

Publications:

Behrstock, Robert A. 2008. *Dragonflies and Damselflies of the Border Southwest.* Tuscon, AZ: Rio Neuvo Publishers.

Biggs, Kathy 2009. *Common Dragonflies of California: A Beginner's Pocket Guide.* 2nd ed. Sebastopol, CA: Azalea Creek Publishing.

Cannings, R. 2002. *Introducing the Dragonflies and Damselflies of British Columbia and the Yukon.* Victoria, BC: Royal British Columbia Museum.

Corbet, P. 1999. *Dragonflies: Behavior and Ecology of Odonata*. Ithaca, NY: Cornell University Press.

Dicken, Samuel N. 1965. *Oregon Geography.* Ann Arbor, MI: Edwards Brothers.

Dunkle, Sydney W. 2000. *Dragonflies through Binoculars: A Field Guide to Dragonflies of North America.* New York, NY: Oxford University Press.

Federal Research Natural Areas in Oregon and Washington: A Guidebook for Scientists and Educators. 1972. Portland, OR: Pacific Northwest Forest and Range Experiment Station.

Gordon, Steve and Cary Kerst. 2005. *Dragonflies and Damselflies of the Willamette Valley, Oregon: A Beginner's Guide.* Eugene, OR: CraneDance Publications.

—2006. "2005 *Aeshna* Blitz— The Best Ever." *Argia* 17, no. 4.

Houck, Michael C. and M. J. Cody. 2000. *Wild in the City: A Guide to Portland's Natural Areas.* Portland, OR: Oregon Historical Society Press.

Johnson, Jim. 2006. "Thoughts on *Lestes stultus*—Is it a Valid Species?" *Argia* 18, no. 2 (August):10-11.

Johnson, Jim and Steve Valley. 2005. "The Odonata of Oregon." *Bulletin of American Odonatology* 8, no. 4, (January): 100–122. (This issue is devoted to Oregon Odonates with status, range maps, and habitat preferences, as well as adult flight range dates.)

Jones, Colin D., Andrea Kingsley, Peter Burke, and Matt Holder. 2008. *Field Guide to the Dragonflies and Damselflies of Algonquin Provincial Park and Surrounding Area.* Whitney, Ontario, Canada: Friends of Algonquin Park.

Lam, Ed. 2004. *Damselflies of the Northeast: A Guide to Species of Eastern Canada and the Northeastern United States.* Forest Hills, NY: Biodiversity Books.

Loy, William G., Stuart Allen, Aileen R. Buckley, James E. Meacham. 2001. *Atlas Of Oregon.* 2nd ed. Eugene, OR: University of Oregon Press.

Manolis, Tim. 2003. *Dragonflies and Damselflies of California.* California Natural History Guide No. 72. Berkeley, CA: University of California Press. (This book covers 81 of Oregon's 91 species with excellent descriptions and illustrations.)

Montgomery, B. Elwood. 1972. "Why Snakefeeder? Why Dragonfly? Some Random Observations on Etymological Entomology." Proceedings of the Indiana Academy of Science: 235–241.

Needham, J., M. Westfall, and M. May. 2000. *Dragonflies of North America.* Gainesville, FL: Scientific Publishers.

Oregon Atlas & Gazetteer. 1991. Freeport, Maine: DeLorme Mapping.

Orr, Elizabeth L., William N. Orr, Ewart M. Baldwin. 1992. *Geology of Oregon.* 4th ed. Dubuque, IA: Dendall/Hunt Publishing Co.

Paulson, Dennis R. and Sydney W. Dunkle. 2009. *A Checklist of North American Odonata, Including English Name, Etymology, Type Location, and Distribution.*

Paulson, D. R. 1983. A new species of dragonfly, *Gomphus* (*Gomphurus*) *lynnae* spec. nov., from the Yakima River, Washington, with notes on pruinosity in Gomphidae (Anisoptera). *Odonatologica* 12: 59–70.

Paulson, Dennis. 1999. *Dragonflies of Washington.* Seattle, WA: Seattle Audubon Society.

Paulson, Dennis. 2009. *Dragonflies and Damselflies of the West.* Princeton, New Jersey: Princeton University Press.

State of Oregon. 2009. *Oregon Blue Book.* Portland, Oregon.

Valley, S. 1993. DSA Meeting in Bend, Oregon. *Argia.* 5(2): 3–6.

Westfall, M., and M. May. 2006. *Damselflies of North America.* Gainesville, FL: Scientific Publishers.

Internet Sites:

The California and Southwest Dragonfly sites by Kathy Biggs. "Dragonflies of California (Odonata)," http://www.sonic.net/dragonfly. "Dragonflies of the Southwest (Odonata)," http://southwestdragonflies.net/

Jim Johnson's site has excellent photos of Northwest species, but also photos from exotic places. "Odonata of the Northwest (Oregon and Washington) and Beyond: Photos of Dragonflies and Damselflies by Jim Johnson," http://odonata.bogfoot.net

The Washington Dragonfly Biodiversity site by Dr. Dennis Paulson from the University of Puget Sound's Slater Museum of Natural History. "Dragonflies—University of Puget Sound," http://www.ups.edu/x5666.xml

The Odonata Central site by Dr. John Abbott of the University of Texas in Austin has excellent range maps, checklists, and photographs with coverage of North America, including Mexico. It is also the official web site of the Dragonfly Society of the Americas. "OdonataCentral," http://odonatacentral.org/

Internet Discussion Groups:

General Odonata, Dr. Dennis Paulson

http://mailweb.ups.edu/mailman/listinfo/odonata-l

Pacific Northwest Odonata, Jim Johnson

http://pets.groups.yahoo.com/group/nw_odonata/

Odonata Association:

Dragonfly Society of the Americas, c/o Jerrell Daigle, 2067 Little River Lane, Tallahassee, FL 32311. Membership includes *Argia* (a news journal) and an additional fee covers the *Bulletin of American Odonatology*. http://www.odonatacentral.org/index.php/PageAction.get/name/DSA_Membership

Equipment:

This company's online catalog provides a wide range of nets, magnifying lens, and other entomology products: BioQuip Products, 2321 Gladwick St., Rancho Dominguez, CA 90220. Phone: 310-667-8800. http://www.bioquip.com

Index

Definitions, illustrations, and descriptions are shown in regular type; photographs in bold. Larval photographs accompany each family description [e.g., Darners (larva) 72, **(72)**.] Some species have extra photographs shown in the "Wonders of Dragonflies," and they are noted with a prefix "W" (e.g., Common Green Darner 78, **79**, **W289**.)